EARTH RETENTION SYSTEMS

EARTH RETENTION SYSTEMS

Alan Macnab P.Eng.

McGRAW-HILL
New York   Chicago   San Francisco   Lisbon   London
Madrid   Mexico City   Milan   New Delhi   San Juan
Seoul   Singapore   Sydney   Toronto

Catalog-in-Publication Data is on file with the Library of Congress.

McGraw-Hill
A Division of The McGraw·Hill Companies

Copyright © 2002 by The McGraw-Hill Companies, Inc. All rights reserved. Printed in the United States of America. Except as permitted under the United States Copyright Act of 1976, no part of this publication may be reproduced or distributed in any form or by any means, or stored in a data base or retrieval system, without the prior written permission of the publisher.

1 2 3 4 5 6 7 8 9 0 DOC/DOC 0 7 6 5 4 3 2

ISBN 0-07-137331-4

The sponsoring editor for this book was Larry S. Hager and the production supervisor was Sherri Souffrance. It was set in Times Roman by Lone Wolf Enterprises, Ltd.

Printed and bound by R. R. Donnelley & Sons Company.

This book is printed on recycled, acid-free paper containing a minimum of 50% recycled, de-inked fiber.

McGraw-Hill books are available at special quantity discounts to use as premiums and sales promotions, or for use in corporate training programs. For more information, please write to the Director of Special Sales, McGraw-Hill, Professional Publishing, Two Penn Plaza, New York, NY 10121-2298. Or contact your local bookstore.

Information contained in this work has been obtained by The McGraw-Hill Companies, Inc. ("McGraw-Hill") from sources believed to be reliable. However, neither McGraw-Hill nor its authors guarantee the accuracy or completeness of any information published herein, and neither McGraw-Hill nor its authors shall be responsible for any errors, omissions, or damages arising out of use of this information. This work is published with the understanding that McGraw-Hill and its authors are supplying information but are not attempting to render engineering or other professional services. If such services are required, the assistance of an appropriate professional should be sought.

CONTENTS

Preface vii
Acknowledgments ix
About the Author xi

Chapter 1 Introduction	1
Chapter 2 Types of Excavations	9
Chapter 3 Types of Shoring Systems	23
Chapter 4 Lateral Support	95
Chapter 5 Facing	187
Chapter 6 Shoring Uses	219
Chapter 7 Investigations	249
Chapter 8 Engineering Properties of Retained Soils	259
Chapter 9 Forces on Walls	267
Chapter 10 Failure Modes of Shoring	291
Chapter 11 Design Methods	303
Chapter 12 Ground Water Control	325

Chapter 13 Installation Equipment and Techniques	345
Chapter 14 Load Testing of Anchors	393
Chapter 15 Monitoring	407
Chapter 16 OSHA Regulations (Standards—29CFR)	419
Chapter 17 Computer Design	495
Chapter 18 Tables	499
Chapter 19 Bibliography	521
Glossary	523
Index	527

PREFACE

When I was first approached to write this book, I asked Larry Hager of McGraw-Hill why he had focused on a topic which, to the general public, is not a high visibility issue. He replied that he had witnessed the death of a laborer, quite close to his house, who was killed in a cave-in while hooking up a new tank being installed at a nearby gas station. As a result of that accident, Larry felt that there should be a reference manual so that this type of event did not recur.

When I pointed out to Larry that the subject was covered to some extent by the OSHA regulations, he acknowledged that a lot of the information necessary was there. However, he felt that the information was not in a format which was readily accessible for the general public. Out of his concern was born this project.

For those of us involved in the shoring business, the dangers of excavations are known. But it is evident that those dangers are not apparent to the vast majority of the public today. We see improperly shored or un-shored excavations every day. I once had a Washington State Department of Labor and Industries Officer explain to me that over 50 percent of the citations he issued were a result of something he saw published in the newspaper which lead him to believe that something unsafe was occurring, prompting him to investigate further.

It is hoped that this text will be made available to every person who has authority over an excavation. The text is not intended to be a how to, nor a definitive treatise on design, but rather an introduction for persons of all technical skill levels to the basic shoring procedures which are available today.

Unlike other aspects of civil engineering, such as the design and construction of large commercial structures, the practice of shoring is almost without standards or codes. The methods of shoring have been developed largely by individuals functioning within small privately held contracting companies. These companies are run by individuals who are just that—individuals. They are intensely competitive and extremely entrepreneurial. Their companies survive and thrive because of their innovations. It should not surprise the reader that the practice of shoring is therefore very regional and quite without standardization.

This text will outline some of the methods used, and some of the more common techniques practiced. It is not intended to be, nor could it ever be, a defining document for the purposes of specifying shoring methods. It is hoped that the readers will gain a level of understanding such that they may be able to comprehend the basis for the design and construction of the shoring proposed for a particular project. The shoring of any given excavation should be entrusted to individuals and companies who are experienced in this work. Their diligence is the greatest safeguard of the public and those workers directly involved in the project.

It is my sincere hope that this book will raise the consciousness of those involved in excavations, be it through design, permitting, construction, or procurement, so that they will ensure that proper procedures are implemented and experienced personnel are engaged. In this day and age, when the knowledge and skill required to be safe is easily obtainable, it is simply inexcusable for us to continue to read news articles about cave-in accidents and deaths in underground construction.

ACKNOWLEDGMENTS

In writing this book, I had help from a lot of friends. This book involved a lot of support and input from people whom I have been close to in the business world for many years.

Those who helped me with job photos and drawing details include Bill Grady of KLB Construction, Brian Isherwood and Nadir Ansari of Brian Isherwood and Associates, Troy Adams of Skyline Steel, Scot McKellar of Hurlen Construction, Peter McDonald, and Bill Starke of Deep Foundations Contractors, Vince Jue of Champion Equipment, Ernie Brandl of Schnabel Foundation, Charlie Griffiths of CT Engineering, Stan McAllister of City Transfer, Horst Aschenbroich of Con-Tech Systems, Bruce Jensen of Williams Form Engineering, Robert Fisher of Dywidag Systems, Bill Warfield of Ingersoll Rand, Mike Weekes and Gary Berg of Stoneway Concrete, Don Morin of DMI Drilling, Jim Su of CivilTech, Alan Rasband and John Roe of Malcolm Drilling, John Monroe of Watson Inc, Matthias Heichel of ABI, Inc, Dan Fruhling of Fruhling Excavation, Ron Boscola of Murray Franklin, Tom Gurtowski of Shannon & Wilson, Ben Dutton of Equipment Corporation of America, Arnie Carson of KPFF Consulting Engineers, Michael Leffer of Hans Leffer GmbH, Gary Schnee of Seaport Steel, Ed Sneffen of Golder Associates, Tony Barley of SBMA, Shannon Creson of Drill Tech Drilling and Shoring, and Bob Federighi and Matt Partain of Condon Johnson & Associates.

Colin Cox and Eric Lindquist of Golder Associates were tireless in their development of illustrations. I would also like to thank Pat Sievertsen and Leo Stapleton of Condon Johnson & Associates for their assistance with photo compilation. Skip Dickman of Dickman Hines Lumber provided insights on lumber grading and treatment, Joel Moskowitz of Mueser Rutledge Consulting Engineers gave me a view of the history of earth retention in New York City, and Bob Plum of Golder Associates was in invaluable resource for locating reference materials.

Scot Litke of the ADSC–The International Association of Foundation Drilling, and Roger Woodson of Lone Wolf Enterprises get high marks for encouraging me throughout the process.

A special thank you to David Cotton of Golder Associates for his assistance, not only with technical recommendations and arranging Golder's collaboration in providing illustrations, but also for taking the time to proofread chapters of the book and keep me going when times got tough.

The book would not have been possible had it not been for two people who have had a profound impact on my career and life. Dr. Hugh Peacock, my structures professor at the University of Western Ontario, rescued me from a life of civil service boredom and injected me into the world of construction. Once I was pressed into service in construction, Bill Lardner of Deep Foundation Contractors took the roll of mentor in my life, and showed me the joy of construction in the underground.

And last but certainly not least, to my wife Anne, my proofreader and supporter who never batted an eye when the book took over my life—because she knew it made me happy.

And to my golf partners, I assure you that in the future, I will not lose as much money to you on the golf course as I did this year because I will now be a little more focused on the game at hand.

ABOUT THE AUTHOR

Alan Macnab's entire career has been spent with earth retention construction companies involved in both design-build and hard bid construction. His roles in construction have included work in operations, supervision, estimating, marketing, and claims, including executive positions in both Canada and the United States. He has held positions in a number of industry associations, including the Presidency of the ADSC–The International Association of Foundation Drilling (1990-1991) where he continues to chair the Standards and Specifications Committee. He is currently Vice President, President-Elect of the Geo-Institute of the American Society of Civil Engineers. Macnab received his B.E. Sc. from the University of Western Ontario and continued there with graduate studies. He is licensed as a professional engineer in Canada.

Macnab has been involved in the construction of earth retention and deep foundation projects since 1973. These projects have been implemented throughout North America, including eastern and western Canada, the northeastern states, Rocky Mountain states, and West Coast. He has traveled extensively, reviewing construction techniques in North America, Europe, Asia, and the South Pacific. He is a frequent speaker on issues involved in these specialties to engineering, academic, and supervisory personnel and has written extensively on these topics.

CHAPTER 1
INTRODUCTION

The body of work called variously Earth Retention, or Shoring, or Geo Support, or Sheeting has historical roots. Earth Retention, as we know it today, has been the amalgamation of construction technologies, equipment innovations and engineering analyses borrowed from many other disciplines. The real coalescing of these roots into a distinct discipline did not occur until well into the latter part of the 20th century, but now represents literally billions of dollars of work annually in the United States alone.

Earth Retention systems are created by a contractor drilling, driving and excavating, and an engineer investigating, analyzing, predicting, measuring and confirming. The innovations of the past have come together in the 20th century to form a critical mass which has evolved into the shoring industry as we know it today.

PILING AND PILE DRIVING

We may never know how those first timber piles, found in Swiss lakes, which supported stilt type houses from the period of 3000 BC were installed, but we assume that in some manner they were driven into the ground. Pile driving was born. We do know, however, that the Greeks were driving piles in 1000 BC and that later the Romans also performed pile driving. The driving of timber piles continued through the ages. Just when piles were first used for their lateral capacity as a retaining wall is open to conjecture. It may have been for fortifications where parallel lines of vertical timber piles were installed and fill was placed between them to create breastworks.

Advancements in shoring occurred when the timber piles used as retaining walls were replaced by squared timbers with tongue and groove joints for tighter fit. Later, several timbers were ganged together to form Wakefield sheeting. This form of tight sheeting survived until the early 20th century, when the patenting of steel sheet piling rendered timber sheeting obsolete. Steel sheet piling wasn't really put into extensive use until after World War I.

Piling was driven with drop hammers up until the introduction of air and steam hammers in 1845 in this country. The driving of steel sections appears to have begun around 1880, about the same time as the introduction of the shoring system called the "Berlin Method," named for the city of its origin. This was the origination of soldier pile and lagging. Soldier piles were primarily installed by impact driving in the U.S. until well into the 1950s. The first drilled soldier piles in Toronto were not installed until the early 1960s.

While the use of the vibro hammer was pioneered by the Russians in the 1920s, it really didn't reach the U.S. market until the 1950s. Originally, "vibros" were chain driven, electric devices. These were modified and redesigned in the U.S. in 1969 with the revolutionary introduction of the hydraulic vibration hammers which are in use today. Most sheet piling is currently driven this way. In a parallel development, the sonic hammer or bodine hammer was introduced in the early 1960s. While showing great promise, the sonic hammer never gathered widespread support and so remains one of those good ideas which never fulfilled its advance billing.

The diesel piledriving hammer, invented in Germany in the late 1930s, came into use in the U.S. in the 1950s. It, together with the vibro hammer, largely replaced air and steam hammers for driving soldier piles.

DRILLING

A Chinese building code of 1103 AD records the use of excavated shafts as a form of foundation and it is generally accepted that all civilizations have excavated holes as a way of forming foundations. Hand excavated pneumatic caissons were first used in France in 1839, and in the U.S. in 1852 for the excavation of bridge piers. The first recorded use of the pneumatic caisson for a building in the U.S. was 1893 in New York City. In a parallel development, a formalized process of foundation production was instigated in the 1880s called the Gow Caisson or Chicago Caisson. This was a hand-dug hole, large enough for a man to enter, which was shored as excavation progressed. Instead of excavating to water and calling it a well as man had done since the dawn of time, pit miners sank shafts to a good bearing layer and called it a caisson.

Just when this hand excavation method was first used for underpinning is open to question, but there is recorded evidence of underpinning including temporary shoring of a retaining wall in France in the 1690s. With the onset of extensive building construction, which included significant substructures, in the urban cities of America in the 1880s, it can't have been long afterward that underpinning was required and the hand-dug underpinning pit was born.

The mechanical drilling of shafts began in Texas in the 1920s when horse-driven augered shafts were installed to overcome expansive soils which bedevilled the local builders. The early 1930s brought the first power-driven auger rigs and the drilled shaft industry took off. When the first soldier pile was installed by drilled methods in lieu of driving is open to argument, but by the late 1950s soldier piling, by the drilled and placed method, was becoming popular.

SOILS INVESTIGATION

The first recorded Standard Penetration Test (SPT), which of course was not standard at the time, was performed in a wash bored test shaft in 1902. By 1914, it had become standardized and is one of the many *in situ* tests which engineers now use to evaluate the soils in which excavations are made. A variety of cones, penetrometers, pressuremeters, and piezometers in use today all provide the input values for the analysis used for design of shoring.

ANALYSIS AND DESIGN

At the risk of leaving out significant parties, the shoring engineer can look back over a few seminal points in history to identify the basis of the design methods we use today. In 1770s Coulomb produced his theories on design for retaining walls and many are still in use. By 1857, Rankine had developed Earth Pressure Theory based on active and passive pressures. Some of his diagrams are still used to design cantilever and single level of bracing shoring.

It has been noted that by 1906 a Mr. J.C. Meems was writing about earth pressures in trenches and, although his work has largely been disregarded, it indicates that the profession was looking for ways to rationally design excavation support systems.

In 1943, Dr. Karl Terzaghi wrote papers on Wedge Theory and, as a result of work with strut loads on deep cuts in the Chicago Subway, he and Dr. Ralph Peck developed the diagrams that are used today for multi level bracing systems. With the advent of reinforced earth (1950s) and soil nailing (1970s), different methods of analysis were developed and engineers now often use Limit Equilibrium methods to solve their earth retention problems.

MEASUREMENT

The first recorded installation of steel struts in a shored excavation was 1926. Prior to that time, timber strutting had been used. It is a credit to the thoughtfulness of those early engineers, that those struts on the first job were instrumented with strain gauges.

Movement of excavations, which originally were measured against fixed baselines with levels and transits are, with the invention in the U.S. of the slope indicator casings in 1958, being monitored with much more accuracy. Global Positioning (GPS) methods, which really found their way into construction in the 1990s, have eased the problems involving measurement which used to require careful maintenance of fixed monitoring points.

SLURRY

The use of slurries to maintain stability in an otherwise unstable hole was born in the petroleum industry in 1914 when it was found that deep holes could be stabilized with slurries of natural material. The use of bentonite was originated by the oil well drilling industry 1929 and adapted for use in the drilled shaft industry in the 1950s. This technique, together with the driving of casing, made commonplace by the vibro hammer, has massively influenced the expansion of drilling in materials otherwise considered to be inappropriate for shaft excavation.

The first slurry trench cutoff walls for ground water control were installed in the U.S. in 1948 and the first structural slurry walls were constructed in Italy in 1950. Structural slurry walls did not appear in the U.S. until 1962. The hydrofraise excavation methods were derived in Europe in 1960 and the method arrived in U.S. in 1970.

TIEBACKS

While anchorages were recorded in Europe in 1874 and tiedowns made up of driven piles and screw piles are recorded in this country as early as 1902, the first drilled, post tensioned tiebacks were actually installed in Algeria in 1934. Drilled anchor technology for permanent anchors did not reach Europe until the 1950s and the U.S. until the 1960s, although there is some record of the use of driven beam tiebacks in the 1950s in the U.S.

Prior to this time, lateral earth pressures in deep cuts were restrained with either struts or rakers. In fact, until the end of World War II, most internal bracing utilized timber.

The first tensioned mechanical screw anchors were installed in U.S. in 1963, and drilled and belled anchors were installed in Toronto in 1965. In the early1960s, Europeans began investigating the development of frictional capacity in soil anchors and by the end of that decade regrouting techniques for anchors were being used in Europe.

Driven casing methods, for the purpose of installing anchors, were first introduced in 1970 in Europe and soon thereafter in the U.S. Today, the majority of anchors are installed with some form of what is now called duplex drilling which involves advancing a casing simultaneously with the drill bit.

With this onslaught of drilled anchors came specifications and consensus documents. The Post Tensioning Institute (PTI) issued its first recommendations for soil and rock anchors in 1976 and in 1991 the International Association of Foundation Drilling (ADSC) recognized the work of anchored earth retention as being within the scope of its responsibilities.

DEWATERING

During the 1890s, the work of installing deep foundations for major buildings in waterbearing sands in New York was being performed by Pneumatic Caisson methods. In this method, hand excavation was carried out under air pressure within an enclosed box. This is not to say that these were the first sunken caissons. Sunken masonry caissons are recorded as early as 1204 AD in Egypt. The first evidence of an excavation where an attempt was made at keeping the excavation dry, which we would now call the cofferdam method, was recorded in 1753 in France. In 1768 an unwatering project was attempted utilizing an undershot waterwheel to develop pumping power. The first real attempts at deep dewatering were not made in the U.S. until 1927, and the technique really didn't become commonplace until after World War II when a tremendous building boom enveloped the country.

SECANT WALLS

The first use of secant walls is recorded in the 1920s in Europe but it was not until 1950 in the U.S. when continuous pile walls, as they were called, were installed. The 1970s brought forth the introduction, in Japan, of soil/cement mixing technology. Methods were devised to perform mixing to considerable depths. This method, now referred to as the Deep Mixed Method (DMM), was first introduced into the U.S. in the early 1980s, but it wasn't until 1986 that a large commercial application was performed. DMM is now commonly used to create deep secant pile walls for earth retention purposes. This technology has a number of uses as a ground improvement tool which bodes well for its continued use. Soil mixing relies on a knowledge of rheology gained from the grouting industry to assist in its many applications.

SOIL NAILING

Although the Romans appeared to be exercising a form of soil nailing when they drove timber piles for slope stability improvement, soil nailing as we know it was first introduced in France in 1972 and in U.S. the in 1976.

MICROPILES

Micropiling was first developed in Italy in the early 1950s where it was used as a method to repair war damage. The first North American application of micropiling came in Canada in 1971 and micropiles were installed followed soon thereafter in the U.S. in 1973.

COMPUTERS

Not only has the proliferation of computers in the 1980s and 1990s changed the face of data logging when measuring earth retention performance, every engineer and contractor has one on his/her desk. The tool to deal quickly with the tiresome iterative solutions so inherent in moment calculations, or to optimize strut configurations, or to solve limit equilibrium problems is at the engineer's finger tips today. Engineering calculations are far less burdensome than even 25 years ago.

THE INDUSTRY

With the onset of the building boom in the 1880s in the U.S. came the formation of specialized foundation companies which performed subgrade works. Because the solutions of the time involved a more integrated relationship between temporary works and the completed structure, these companies performed all the foundation work. It would have been impossible to separate the work of excavating a pneumatic caisson from the subsequent construction of the footing or wall within the caisson. With this specialization came the creation in the early 1900s of specialty engineering firms who performed soils investigations and foundation designs.

As the shoring industry developed, the shoring schemes became less integrated with permanent construction. In North America, the work of soldier pile and lagging was initially performed by general contractors who utilized the services of structural engineers for design and piling contractors as subcontractors for the installation of the driven soldier piles. However, by 1960 the practice of shoring had advanced to the point where engineers with specialized skills in earth retention design could support themselves on a steady diet of this type of work. At the same time, specialty contractors staffed with civil engineers had taken over the construction of complete shoring systems and were offering those systems on a design-build basis.

As one can readily see, the pieces of the puzzle that go together to form the knowledge and skill base of the shoring engineer and contractor come from many roots. Some are home grown, while many have been taken from foreign lands and different industries. The 21st century is bringing added technological change in the form of methods designed to reduce the amount of open cut necessary to construct today's infrastructure. These methods include microtunnelling, New Austrian Tunnelling Methods (NATM), trenchless technology and directional drilling. In spite of this, there can be no question that the amount of earth retention work will continue to expand. Newer innovations and additions will continue to change and refine the part of earth retention in underground construction, but the earth retention industry will remain a dynamic industry peopled by innovators.

CHAPTER 2
TYPES OF EXCAVATIONS

While the ultimate purpose of the excavation does not define the shoring which may be required to protect it, the type of shoring used often defines the type of excavation which may be undertaken. The life cycle of an excavation has a large input into the decision as to which method is used. Excavations which will only be open for a very short period of time are often shored with very different methods than might be used for longer periods of time. In fact the length of time that an excavation is open may determine if it is shored at all.

Excavations are shored for a variety of reasons. They may be shored to limit the amount of overexcavation required when sloping the sides of the cut. They may be shored to protect the personnel who enter and work within the excavation. Shoring may be placed to protect adjacent property such as buildings, utilities or property for which no easement is available. Shoring also may be installed to minimize the excavation and therefore maximize the usable property around the excavation. In doing so, close access for hoisting into the excavation and storage of materials slated to be used in the excavation can be enhanced.

2.1 TRENCHES

Trenches are long narrow excavations, usually deeper than their width, which are intended to be open for a brief period of time. Trench excavations are often made for the installation of utilities (see Figure 2.1), but may also be used to install water cutoff barriers or drainage elements.

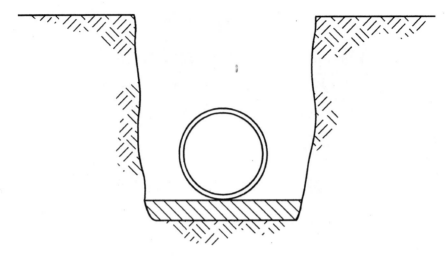

FIGURE 2.1 Typical trench section.

Utilities such as telephone, gas or electricity which do not rely on a hydraulic gradient are usually quite shallow and the grade of these utilities will often follow the surface profile of the adjacent ground. Shoring requirements for these utilities are usually restricted to vault structures where valving or connections are made. The depth of the trench is often shallow enough (see Chapter 16; OSHA Regulations) to permit personnel access without shoring. Connections, which need to be made to discrete pieces of pipe or cable, are often made prior to lowering the line into the trench, thereby eliminating the risk associated with placing personnel in the trench.

Sewers are usually installed in trenches but, contrary to the trenches used for the previously mentioned utilities, these tend to be quite deep. Sewers generally are based on gravity flow principles and this tends to drive the sewer ever deeper into the ground. Sewer construction is made up of distinct pieces of pipe which must be spliced or coupled, generally by personnel in the trench. In an urban environment, the opportunity to lay back the side walls of the trench is usually restricted and as a result, some form of shoring is usually required. Chapter 16 of this book—OSHA Regulations 29 CFR 1926 Subpart P—Excavations, details requirements for excavations not supervised by an engineer. Appendix B of these regulations details sloping requirements for trench sidewalls.

Water lines are pressurized systems and as such do not need to follow gradients the way gravity lines do. They tend, however, to be deeper than electric and gas in order to provide protection against freeze/thaw problems and disruption or damage from later adjacent excavation. Similar to sewer lines, water lines are usually spliced in the excavation and shoring must be provided to protect personnel (see Figure 2.2).

FIGURE 2.2 Shored trench with strutted bracing. *(Courtesy of KLB Construction, Inc. Mukilteo, WA)*

Trenches are often excavated in waterbearing materials. These soils may flow and cause disturbance to adjacent property if excavation continues without dealing with the water. When dewatering is not possible or not economic, the method of trench shoring must prevent the uncontrolled flow of soils into the excavation. Subsequent chapters will deal with water cutoff and dewatering methods.

What separates trenching from other forms of excavation is the length of time that the excavation must remain open. While trenches can be very long (up to several miles), there is rarely any need to maintain the entire length of the trench in an open state at any one time. In fact, only a very short piece of trench must remain open at any given time. In cases where the purpose of the trench is to hold a rigid pipe, the length of trench required at any given time would be that length required to prepare the trench bottom with pipe bedding and place one length of pipe together with the distance required to splice that piece of pipe to its predecessor (see Figure 2.3).

In cases where the trench is for the placement of flexible elements such as cable or flexible pipe, splicing can be done above ground and the trench length is limited to that distance required to accommodate the curvature of the utility together with a suitable length to handle any encasement such as concrete which may be required.

Trenches tend to be shallow when compared to other types of excavation. In practice, other methods of installation of utilities such as Tunneling or Trenchless Technologies become more cost effective when the depth proposed is greater than about 40 feet (12 m).

Because the length of trench required at any given time is very small compared to its overall length, shoring systems which emphasize easy reuse, speed of installation and adaptability to a variety of soil conditions are the most appropriate. These include sheet piling (Chapter 3.1), trench boxes, (Chapter 3.2), and driven soldier pile and "road" plate systems (Chapter 3.5 and 5.1.3). Valving stations or vaults for connections are often shored utilizing timber shoring (Chapter 3.3) or lightweight shoring (Chapter 3.4)

A specialized form of trenching for the installation of water cutoff barriers is made stable by the introduction of slurry (either mineral or polymer) into the excavation. The viscosity of the slurry and lateral pressure exerted by its weight aid in holding the trench open. This type of trenching has been known to reach depths of 80 feet (24 m).

2.2 FOUNDATIONS

Excavations to remove soil for the purpose of installing foundations for structures, buildings or retaining walls are often similar in nature. Soils are removed in order to uncover soil or rock of suitable bearing capacity and to permit placement of those portions of the construction which are designed to be placed below ground level. These excavations are often referred to as mass and structural excavations.

When dealing with the construction of buildings, mass excavation refers to the excavation required to remove soil down to the underside of the lowest basement slab level including soil removed to permit under-slab granular placement. Structural excavation is that excavation required to further deepen the site locally as required for individual footings.

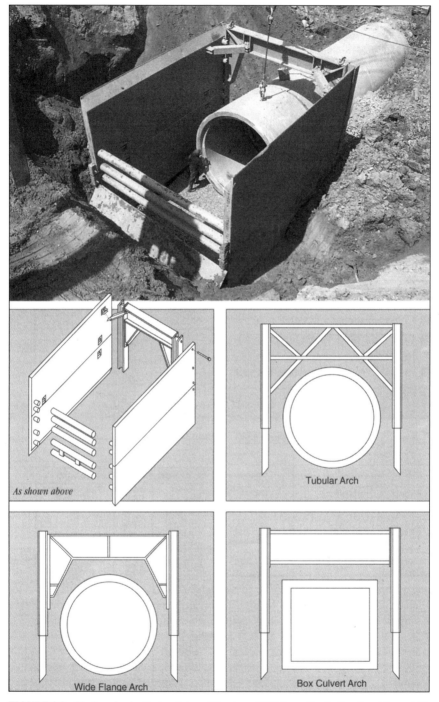

FIGURE 2.3 Placing pipe inside trench box. *(Courtesy of Efficiency Production, Inc. Mason, MI)*

In the case of bridge abutments and piers or retaining walls, structure excavation usually refers to the excavation of soils which are found within the limits of the finished structure plus a small allowance for formwork. All other excavation, such as over-excavation in lieu of shoring, to access the structure excavation is called mass excavation.

Foundation excavations, when complete, must be relatively dry and expose undisturbed soils or rock of sufficient bearing capacity to meet the design requirements of the structural engineer. Once this is done, the excavation must remain in this stable condition until reinforcing steel and concrete are placed for the footings and any walls which may be required to bring the structure to grade. Figures 2.4 and 2.5 outline examples of shoring used to maintain building excavations in an open state.

It is this extended time requirement which separates foundation excavation from trenching and often forms the basis for the decision to shore the excavation. Shoring in these circumstances may be called upon to:

- Protect adjacent utilities and property
- Permit continued access to roadways or property immediately adjacent to the excavation
- Protect personnel within the excavation
- Provide a water barrier
- Prevent basal heave

The situations most likely to be covered by these criteria include

- Buildings and their footings
- Bridge abutments footings and piers footings or pile caps
- Retaining walls
- Pump stations
- Storage tanks
- Substructures for other civil engineering projects such as waste water treatment plants, tunnel portals etc.

The types of shoring most often found on these projects include sheet piling (Chapter 3.1), soldier pile and lagging (Chapter 3.5), soil nailing (Chapter 3.6), secant pile walls (Chapter 3.7), and underpinning (Chapter 3.11). If the completed facility is to include the shoring as an integral part of its structure (see Chapter 6 for further discussion), cylinder pile walls (Chapter 3.8), slurry walls (Chapter 3.9), and micropile walls (Chapter 3.10) might also be used.

FIGURE 2.4 Site excavation—tiedback soldier pile and lagging, Syracuse, N.Y.

FIGURE 2.5 Site excavation—internally braced secant wall shoring, San Francisco, CA. *(Courtesy of Condon-Johnson & Associates, Inc. Oakland, CA)*

2.3 CUT AND COVER

Excavations which are described in this section look a lot like trenches with a few important distinctions. These excavations are usually long and narrow and contain structures which, when completed, are entirely buried below grade. The important distinction is that the excavation must remain open for a considerable period of time in order to construct the structures to be contained within them. Projects such as subways, depressed rail and road beds, and very deep utility excavations not installed by tunneling methods are prime candidates for cut and cover construction (see Figures 2.6 through 2.8).

Cut and cover refers to the process of opening the excavation, placing the structure within the excavated space, and then covering the structure with soil again. The shoring, if required, is generally sheet piling (Chapter 3.1), soldier pile and lagging (Chapter 3.5), soil nailing (Chapter 3.6), or secant walls (Chapter 3.7). Often a cut and cover shored excavation will include temporary decking which rests on the top of the shoring. This decking then permits traffic to use the space overhead while the structure is being constructed below.

2.4. COVER AND CUT

This specialized technique used for shallow tunnels is not practiced nearly as often as cut and cover but has been shown to be effective in situations where the top of the completed structure is too close to the ground surface to permit tunneling. When it is not possible to disrupt traffic for a period of time long enough to construct by cut and cover methods, cover and cut is a viable option. As shown in Figure 2.9:

1. Step 1. Detour the traffic into one half of the right of way while the main vertical elements of the shoring (usually soldier piles or secant piles) are installed. The roof of the intended structure is then cast on grade, bearing directly on top of the piles.
2. Step 2. Traffic is diverted onto the top of the just completed roof structure and the remaining piles and roof structure are constructed in the same manner as step 1.
3. Step 3. Once the entire roof structure is complete, traffic can be returned to its original configuration. Excavation of the tunnel is then commenced from one or both ends of the tunnel using the already completed roof as protection.
4. Step 4. Complete the construction of the tunnel.

   A completed cover and cut tunnel is shown in Figure 2.10.

FIGURE 2.6 Site excavation—tiedback sheet piling, Vancouver, BC. *(Courtesy of ADSC–The International Association of Foundation Drilling, Dallas, TX)*

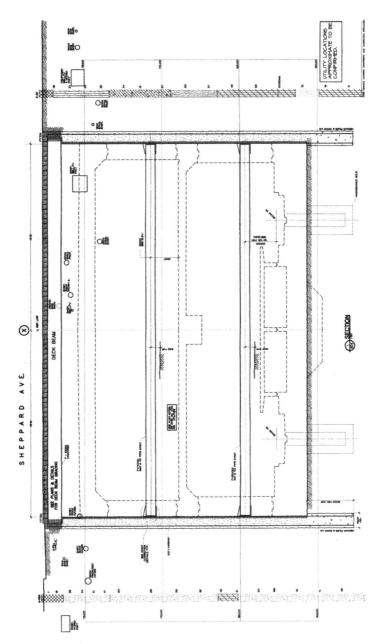

FIGURE 2.7 Subway cut—section, Toronto, Ont. *(Courtesy of Isherwood Associates. Oakville, Ont.)*

FIGURE 2.8 Subway cut, Toronto, Ont. *(Courtesy of Isherwood Associates. Oakville, Ont.)*

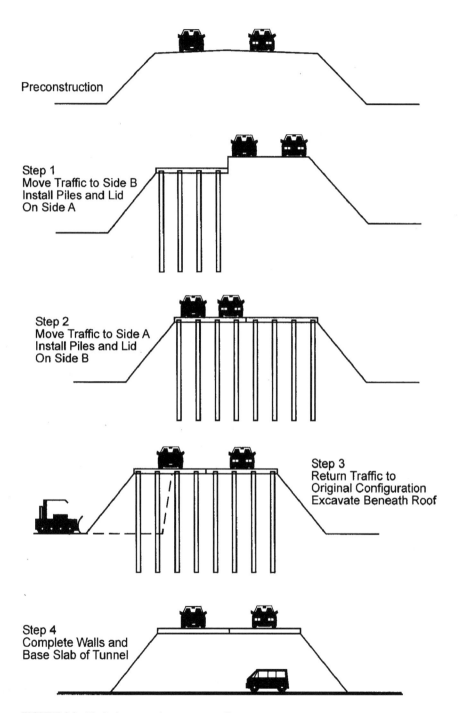

FIGURE 2.9 Typical cover and cut sequence diagram.

FIGURE 2.10 Completed cover and cut, Renton, WA. *(Courtesy of Condon-Johnson & Associates, Inc. Seattle, WA)*

CHAPTER 3
TYPES OF SHORING SYSTEMS

Shoring systems are not standardized within North America or even within a single state of the United States. They are customized installations with the variations depending on local experience, local conditions, availability and cost of materials and the amount of shoring which is performed in a given area. Contractors and engineers, in areas which have large amounts of shoring, tend to develop highly specialized solutions to the particular problems. In areas that have little or no shoring history, the shoring systems tend to be quite textbook in their design and installation.

This chapter will attempt to outline some of the more common techniques used for shoring. It is acknowledged that this chapter cannot possibly cover all the techniques and variations used in North America but will attempt to provide an understanding of shoring systems such that variations, when seen, will not be confusing.

3.1 SHEET PILING

The use of driven sheet piling dates back prior to the development of techniques which permitted the rolling of steel into sheets (see Figure 3.1). Steel sheet piling was patented in the U.S.A. in the 1890s and came into production in the early 1900s. Prior to the introduction of steel sheet piling, when a contractor had to install a shoring system which would retain not only soil pressures, but also water, he might nail three planks together in a staggered fashion (see Figure 3.2) to construct a type of tongue and groove timber sheet which could be driven to excluded running soils. This was called Wakefield Sheeting. Now, with the advent of specialized steel rolling techniques, almost all sheet piling is made from

FIGURE 3.1 Sheet pile bulkhead, Everett, WA. *(Courtesy of ADSC-The International Association of Foundation Drilling, Dallas, TX)*

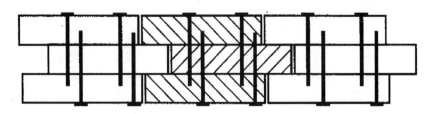

FIGURE 3.2 Timber sheet piling

rolled steel. Some plastic sheeting is being produced for shallow waterfront applications or cutoff barriers but this discussion will limit itself to the types of steel sheet piling.

3.1.1 Sheet Shapes

Sheets can be purchased in a variety of shapes. The most common is a Z shape. These sheets tend to be ¼ to ½ inches (6-12 mm) thick and develop their moment capacity from the depth of the Z. The offset formed by the Z of the sheets is usually 8 to 12 inches (200-300 mm) deep. See Figure 3.3 for typical Z sheet.

In applications where the horizontal stresses are not as high, the sheet of choice will often be a U section. This sheet tends to be wider than the Z section, so fewer joints are required. It is not as deep, however, in its offset so moment capacity is compromised. See Figure 3.4 for a typical U sheet.

Flat sheets (Figure 3.5) are also available but their use in earth retention is severely limited. They have virtually no moment resistance and are used almost exclusively in cellular cofferdams where the sheet is primarily in tension (see Figure 3.6).

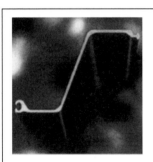

PZ22 AND PZ27 STEEL SHEET PILING

- Cast hot rolled in the U.S.A.
- Available in ASTM A572 Grade 50
- Ball & socket interlock

Dimensions & Properties– Imperial and *Metric*

SECTION	Area in² *cm²*	Width in *mm*	Weight per single lb/ft *kg/m*	Weight per wall lb/ft² *kg/m²*	Moment of inertia in⁴ *cm⁴*	Section modulus per single in³ *cm³*	Section modulus per wall in³/ft *cm³/m*	Surface area per single Total area ft²/ft *m²/m*	Surface area per single Nominal area* ft²/ft *m²/m*
PZ-22	11.86 *76.5*	22 *560*	40.3 *60.0*	22.0 *107.4*	154.7 *6440*	33.1 *542*	18.1 *973*	4.94 *1.51*	4.48 *1.37*
PZ-27	11.91 *76.8*	18 *455*	40.5 *60.3*	27.0 *131.8*	276.3 *11500*	45.3 *742*	30.2 *1620*	4.94 *1.51*	4.48 *1.37*

*Note: Nominal coating area excludes socket interior and ball interlock   Data current as of January 1999

FIGURE 3.3 Sheet pile section—Z sheet. (*Courtesy of Skyline Steel, Inc. Gig Harbor, WA*)

U Sheet Piles

Section Properties

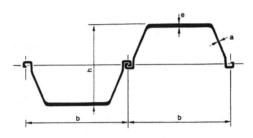

Section	Width b (in)	Height h (in)	Back thickness e (in)	Flange thickness a (in)	Developed perimeter (in/ft of wall)	Section area (in²/ft of wall)	Coating area (ft²/ft of s. pile)	Mass (lbs/ft of pile)	Mass (lbs/ft² of wall)	Section modulus (in³/ft of wall)	Moment of inertia (in⁴/ft of wall)	Radius of gyration r (in)
PU 6	23.62	8.90	.295	.252	28.1	4.54	4.76	30.44	15.5	11.2	49.2	3.30
PU 8	23.62	11.02	.315	.315	29.4	5.48	5.02	36.62	18.6	15.4	85.0	3.94
PU 12	23.62	14.17	.386	.354	31.0	6.61	5.28	44.28	22.5	22.3	157.8	4.89
PU 16	23.62	14.96	.472	.354	32.3	7.51	5.51	50.20	25.6	29.8	223.5	5.45
PU 20	23.62	15.75	.488	.382	34.4	8.50	5.87	56.92	28.9	37.2	292.0	5.86
PU 25	23.62	17.80	.559	.394	35.5	9.45	6.07	63.23	32.2	46.5	413.7	6.62
PU 32	23.62	17.80	.768	.433	35.5	11.48	6.07	77.01	39.1	59.5	529.2	6.79

FIGURE 3.4 Sheet pile section—U sheet. *(Courtesy of Skyline Steel, Inc. Gig Harbor, WA)*

In cases where lateral loads cause extremely high moments in the sheet pile wall, H sheets have been used (see Figures 3.7 and 3.8). These sheets are quite expensive and their use is not common.

3.1.2 Joints

Sheet piles interlock in a number of ways in an attempt to limit the inflow of water and the passage of soil particles through the barrier. Sheets which are hot rolled (rolled directly from billets of steel into their final configuration) have specific joints which are formed during the rolling process. See Figure 3.9 for typical hot rolled joints. Sheets which are cold rolled (rolled into sheet pile shapes from coils of already rolled and finished steel) have a joint similar to Figure 3.10. The cold rolled joint tends not to be as water tight as those found in hot rolled sheets. Cold rolled sheets are usually not used where high water heads might cause excessive infiltration through the joints. These sheets, however, are cheaper

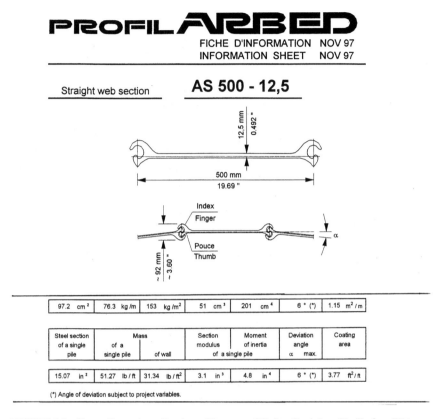

FIGURE 3.5 Sheet pile section—flat sheet. *(Courtesy of Skyline Steel, Inc. Gig Harbor, WA)*

than hot rolled and are used often for temporary earth retention when the potential for water infiltration is low or non-existent.

3.1.3 Uses

Sheet Piling is designed for use in retaining open water situations such as cofferdams in rivers and waterfront retaining walls. It is also extensively used for retaining soils which are below the water table and would flow if excavated. Additionally, sheet piling is very useful in cases of very soft clays which exhibit little or no shear strength or arching potential (Chapter 8.8). Sheet Piling is also used to overcome problems of basal heave (Chapter 9.5) either by embedding the sheet into a stiff, impervious layer, or by installing it deep enough that the flow path around it is sufficient to prevent base instability caused by the upward flow of water.

Sheet piling is also used in cases where small temporary walls are required for structural excavations such as isolated footings. It is often favored for this use as it can be driven quickly and recovered after its use.

FIGURE 3.6 Sheet pile cofferdam for railroad abutment, Cornelius Pass., OR. *(Courtesy of Hurlen, Inc. Seattle, WA)*

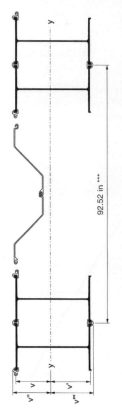

Section	Dimensions				Properties per foot of wall				Mass of combination with intermediary section AZ 13		
	v	v'	v''	v'''	Sectional area	Moment of inertia	*Section modulus	**Section modulus	*lAZ = 60% lHZ* lb/ft²	*lAZ = 80% lHZ* lb/ft²	*lAZ = lHZ* lb/ft²
	in	in	in	in	in²/ft	in⁴/ft	in³/ft	in³/ft			
HZ 575 A	10.35	12.29	11.78	13.73	13.40	1074.0	78.2	87.4	40.55	43.01	45.67
HZ 575 B	10.49	12.31	11.93	13.74	14.14	1173.5	85.4	95.3	43.01	45.67	48.13
HZ 575 C	10.64	12.31	12.07	13.75	15.12	1286.6	93.6	104.5	46.49	48.95	51.41
HZ 575 D	10.70	12.41	12.26	13.96	16.16	1426.8	102.1	114.9	49.97	52.43	55.10
HZ 775 A	14.21	16.30	15.53	17.62	15.70	2199.3	124.8	134.9	48.34	51.00	53.46
HZ 775 B	14.34	16.33	15.66	17.65	16.45	2377.5	134.8	145.6	51.00	53.46	55.91
HZ 775 C	14.43	16.40	15.86	17.84	17.97	2665.6	149.5	162.5	56.12	58.58	61.24
HZ 775 D	14.55	16.44	15.98	17.87	18.71	2848.7	159.4	173.4	58.58	61.24	63.70
HZ 975 A	18.02	20.36	19.34	21.68	17.28	3642.9	168.1	178.9	53.87	56.32	58.78
HZ 975 B	18.16	20.38	19.48	21.70	18.02	3923.0	180.8	192.4	56.32	58.78	61.24
HZ 975 C	18.25	20.45	19.69	21.89	19.86	4418.5	201.9	216.1	62.67	65.13	67.59
HZ 975 D	18.37	20.48	19.81	21.92	20.61	4705.3	214.6	229.7	65.13	67.59	70.05

FIGURE 3.7 Sheet pile section—H sheet. (*Courtesy of Skyline Steel, Inc. Gig Harbor, WA*)

FIGURE 3.8 H pile sheet used for pier construction. *(Courtesy of Skyline Steel, Inc. Gig Harbor, WA)*

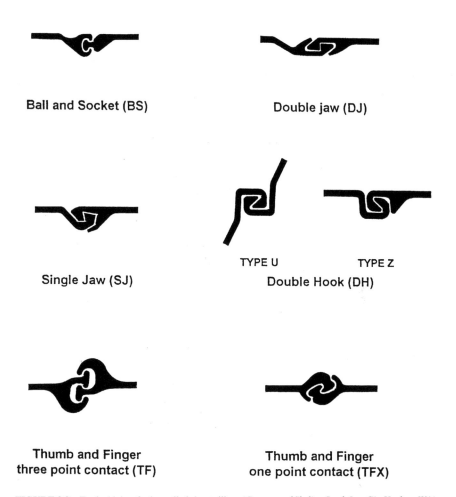

FIGURE 3.9 Typical joints for hot rolled sheet piling. *(Courtesy of Skyline Steel, Inc. Gig Harbor, WA)*

3.2 TRENCH BOXES

Trench Boxes are steel fabrications which are introduced into the trench and dragged along with the excavation. The box configuration is such that it protects personnel in the trench and the work under construction from damage or injury which might be caused by the collapse of the trench sidewalls. Trench boxes are designed to brace the two parallel walls of the trench against each other. The walls of the box are constructed from sheet steel, usually doubled, with a diaphragm between the two sheets to provide structural rigidity.

Casteel Sheet Piling Specifications

Sections	Width b mm	Height h mm	Thickness t[1] mm	Coating Area[2] m²/lin. m	Sectional Area mm²/m	Mass of pile kg/m	Mass of wall kg/m²	Section Modulus 10³mm³/m	Moment of Inertia 10⁶mm⁴/m	Radius of Gyration mm	Sections
CZ67	550	200	5,5	1,46	8 530	36,8	67,0	575	57,50	83	CZ67
CZ72	550	200	6,0	1,46	9 230	39,7	72,0	628	62,80	83	CZ72
CZ84	550	200	7,0	1,46	10 690	46,2	84,0	732	73,25	83	CZ84
CZ95RD	550	200	7,8	1,46	11 820	51,0	92,8	815	81,55	83	CZ95RD
CZ95	550	200	8,0	1,46	12 110	52,3	95,0	835	83,50	83	CZ95
CZ101	550	200	8,5	1,46	12 870	55,6	101,0	887	88,85	83	CZ101
CZ107	550	200	9,0	1,46	13 640	58,9	107,0	940	94,00	83	CZ107
CZ113	550	200	9,5	1,46	14 390	62,0	113,0	993	99,30	83	CZ113
CZ114RD	610	340	8,0	1,80	13 620	65,2	106,9	1 600	272,00	141	CZ114RD
CZ114	610	340	8,5	1,80	14 540	69,7	114,0	1 700	289,00	141	CZ114
CZ128	610	340	9,5	1,80	16 250	77,8	128,0	1 900	323,00	141	CZ128
CZ141	610	340	10,5	1,80	17 950	86,2	141,0	2 100	357,00	141	CZ141
CZ148	610	340	11,0	1,80	18 800	90,3	148,0	2 200	374,00	141	CZ148

FIGURE 3.10 Typical joints for cold rolled sheet piling. (*Courtesy of Skyline Steel, Inc. Gig Harbor, WA*)

The parallel walls of the box are braced apart by adjustable pipe struts. The adjustment permits the use of the box in trenches of different widths. The box is open at both ends. The rear opening permits movement of the box along the trench while allowing passage of the completed pipe utility out the back of the box. The front of the box is open to permit dragging the box forward through unstable ground.

The top of the box is open to permit introduction of pipe bedding or new portions of pipe into the trench. The bottom of the box is open to permit placement of trench bedding directly on the excavated soil.

Trench boxes must not only be wide enough to permit the introduction of the required utility pipe or conduit, together with the specified sidewall backfill cover, but also wide enough to permit passage of the excavator bucket into the box to clean the base of the trench. Trench boxes are dragged forward by the excavator digging the trench. The excavator hooks its bucket behind the leading pipe strut and pulls the box towards itself.

Trench boxes usually extend from the base of the excavation to the original ground, although it is possible to over-excavate the top of the trench in a sloped fashion down to the top of the box (see Figure 3.11).

Boxes come in a variety of lengths and heights. They are usually fabricated by a manufacturer, although some are built by contractors for specific job requirements. Typical size ranges are 24 feet (7.3 m) long by 8-10 feet (2.4-3.0 m) high. Figures 3.12 and 3.13 show typical trench boxes in use today. Trench Boxes can be ganged together to create a box which will protect deeper trenches. Depths of up to 35 feet (11 m) have been shored using this method (see Figure 3.14).

3.3 TIMBERED SHORING

Timbered shoring is probably one of the first methods of shoring ever used. While it becomes quite cumbersome with depth and increased width, it can be a very economical solution for shallow excavations (less than 15 feet (4.5 m)) where the excavation sidewalls are parallel and less than about 12 feet (3.7 m) apart.

The system relies on soil stability to permit the excavation of the pit. The soil must exhibit sufficient standup time to allow placement of bracing sets and vertical timber lagging. A typical timbered excavation would have a bracing frame made of 6 x 6 (150 by 150 mm) or 8 x 8 (200 by 200 mm) timber sitting directly on the bottom of the excavation. Another bracing set would be suspended somewhere close to the original ground elevation excavation. Vertical 2 or 3 inch (50-75 mm) timbers are placed side by side around the perimeter of the bracing sets.

This shoring system can only be used in cases where the proposed structure within the excavation does not need to be constructed of cast-in-place concrete placed directly against the timber shoring. Timber shoring is always intended to be removed. A typical timber shoring arrangement is shown in section in Figure 3.15 while a photo of a timbered excavation site is shown in Figure 3.16. Figure 3.17 shows bracing and sheeting details.

FIGURE 3.11 Sloped cut over trench box in sewer excavation. *(Courtesy of KLB Construction, Inc. Mukilteo, WA)*

TYPES OF SHORING SYSTEMS 35

PXL SIDEWALLS

Model (Ht. x Lg.)	Weight (Lbs.)	Pipe Clearance "C" (In.)*	Shield Capacity (PSF) (Lbs.)	Depth of Cut (Ft.) Soil Type**		
				A	B	C
420 PXL	9,180	22	2,820	113	63	47
424 PXL	10,670	22	2,460	98	55	41
427 PXL	13,200	22	1,980	79	44	33
430 PXL	14,630	22	1,620	65	36	27
432 PXL	15,050	22	1,440	58	32	24
620 PXL	12,910	41	2,280	91	51	38
624 PXL	14,500	41	1,560	62	35	26
627 PXL	16,600	41	1,200	48	27	20
630 PXL	18,060	41	960	38	21	16
632 PXL	19,070	41	840	34	19	14
820 PXL	15,050	65	2,460	98	55	41
824 PXL	17,450	65	1,680	67	37	28
827 PXL	19,400	65	1,260	50	28	21
830 PXL	21,070	65	1,020	41	23	17
832 PXL	22,320	65	900	36	20	15
1020 PXL	18,318	85	2,160	86	48	36
1024 PXL	21,380	85	1,800	72	40	30
1027 PXL	23,700	85	1,380	55	31	23
1030 PXL	25,940	85	1,020	41	23	17

FIGURE 3.12 Typical trench box—Note box being moved by backhoe. *(Courtesy of Efficiency Production, Inc. Mason, MI)*

HT8 SIDEWALLS

Model (Ht. x Lg.)	Weight (Lbs.)	Pipe Clearance "C" (In.)*	Shield Capacity (PSF) (Lbs.)	Depth of Cut (Ft.) Soil Type**		
				A	B	C
416 HT8	5,420	22	2,430	97	54	41
420 HT8	6,760	22	1,710	68	38	29
424 HT8	7,920	22	1,170	47	26	20
427 HT8	8,810	22	970	39	22	16
430 HT8	9,720	22	750	30	17	13
616 HT8	7,160	41	2,430	97	54	41
620 HT8	8,990	41	1,710	68	38	29
624 HT8	10,530	41	1,170	47	26	20
627 HT8	11,530	41	970	39	22	16
630 HT8	13,950	41	750	30	17	13
816 HT8	8,350	65	2,430	97	54	41
820 HT8	10,070	65	1,710	68	38	29
824 HT8	11,800	65	1,170	47	26	20
827 HT8	13,390	65	970	39	22	16
830 HT8	14,460	65	750	30	17	13
1016 HT8	10,550	85	2,100	84	47	35
1020 HT8	13,390	85	1,860	74	41	31
1024 HT8	15,100	85	1,260	50	28	21
1027 HT8	16,800	85	1,050	42	23	18
1030 HT8	17,840	85	750	30	17	13

FIGURE 3.13 Typical trench box—Note placement of precast sections. *(Courtesy of Efficiency Production, Inc. Mason, MI)*

TYPES OF SHORING SYSTEMS

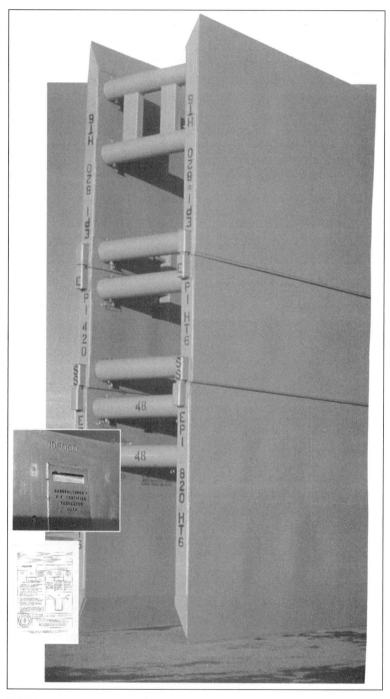

FIGURE 3.14 Stacked boxes. *(Courtesy of Efficiency Production, Inc. Mason, MI)*

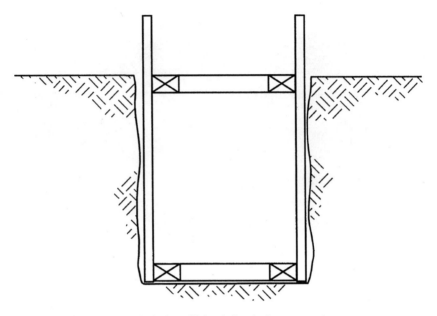

FIGURE 3.15 Typical trench shoring utilizing timber shoring.

This shoring method is light. Each of its component parts can be lifted by hand. Even the bracing sets, when put together, can be easily lifted with the excavator performing the excavation work and the vertical timber can be set by hand. Timber shoring is ideal for:

- Repairs to utilities
- Construction of utility vaults
- Cable splicing
- Later connections of side sewers

Typical timber bracing solutions are included in OSHA regulations 29 CFR 1926 Subpart P, attached in Chapter 16 of this book. These typical solutions allow a contractor to shore an excavation to a depth not to exceed 20 feet (6.1 m) without the supervision of an engineer. This, together with the availability of lumber, ensures that the emergency repairs so often found in the above list can be implemented as soon as a work crew can be brought to the site.

3.4 LIGHTWEIGHT SHORING

Aluminum prefabricated trench shoring is used in applications similar to timber shoring. It is designed to be light and flexible in application. The prefab walls are

FIGURE 3.16 Timber sheeting of utility vault-Seattle, WA.

FIGURE 3.17 Interior of timber sheeted vault with walers in place, Seattle, WA.

constructed of two sheets of aluminum sandwiched over styrofoam for structural rigidity. It is braced by its corner connections or by telescoping pipe struts which can be varied in length to meet job requirements. The pipe struts may or may not be hydraulically adjustable. It is advertised for use to depths of 25 feet (7.6 m). Manufacturer's literature on this shoring solution is shown in Figures 3.18 and 3.19.

3.4.1 Speed Shores

A simple form of shoring for shallow trenches consists of plywood sheets placed against opposing vertical cut faces. The plywood sheets are held apart by hydraulic jacks which are placed horizontally between the sheets. These jacks may be installed one at a time or can be paired up into frames for easy installation (see Figure 3.20).

3.4.2 Shields. Trenching in urban environments brings on its own complications. Often a trench which would seem to lend itself to a trench box cannot be shored in this manner because of a myriad of other utilities crossing the proposed trench. In these situations, utility contractors will often utilize a shield similar to that indicated in Figure 3.21.

FIGURE 3.18 Typical aluminum trench shoring systems. *(Courtesy of Efficiency Production, Inc. Mason, MI)*

Shown with 3" sidewalls and optional arch.

Shown with 5" sidewalls

FIGURE 3.18 *(continued)* Typical aluminum trench shoring systems. *(Courtesy of Efficiency Production, Inc. Mason, MI)*

FIGURE 3.19 Typical lightweight shoring box. (*Courtesy of NES Trench Shoring. Houston, TX*)

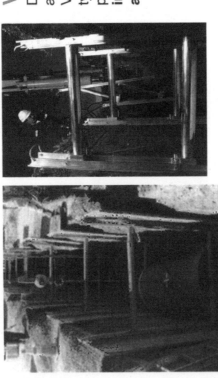

FIGURE 3.20 Speed shores. Note the hydraulic spreaders. *(Courtesy of NES Trench Shoring, Houston, TX)*

The shield consists of series of vertical posts which are maintained in relation to each other by a template at the ground surface. Between the posts, which are strutted apart, trench sidewall protection can be lowered. If an existing utility interferes with a particular panel of the sidewall protection, that portion of the protection can be held above the interfering utility and the portion of the trench wall below the utility can be shored utilizing other means such as timber lagging (Chapter 5).

FIGURE 3.21 Shield system—segments jumped ahead rather than dragged. *(Courtesy of NES Trench Shoring. Houston, TX)*

The shield system consists of a number of panels (usually three) in place at any one time. As work is completed in the first panel, the trench is backfilled and that panel removed and reattached at the end of the third panel to permit further construction of the utility.

3.5 SOLDIER PILE AND LAGGING

Soldier pile and lagging is probably the most common shoring solution for urban construction. Soldier piles are vertical steel elements which define the perimeter of the excavation. Spaced at 6-10 feet (1.8-3.0 m) on center, they stand at attention like soldiers, hence their name. The spaces between the soldier piles are filled with lagging (see Chapter 5). Figure 3.22 details a typical soldier pile and lagging arrangement while Figure 3.23 is a photo of a completed soldier pile and lagging project. Soldier piles can be driven, drilled and concreted, churn drilled, or wet set in soil cement.

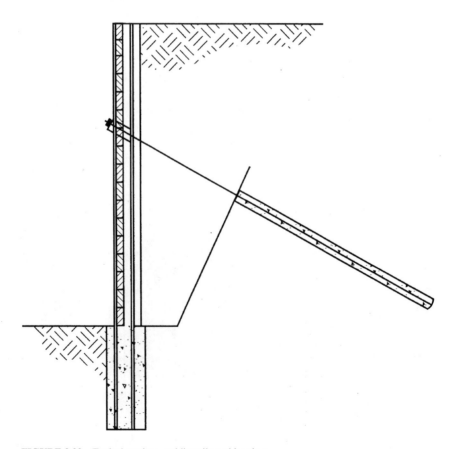

FIGURE 3.22 Typical section—soldier pile and lagging.

FIGURE 3.23 Soldier pile and lagging for slide repair, Allyn, WA. (*Courtesy of Condon-Johnson & Associates, Inc. Seattle, WA*)

3.5.1 Types of Soldier Piling

3.5.1.1 Driven Soldier Piling. When driven, soldier piles are usually H sections although some wide flanged sections are used when the driving stresses are light. Driven piling sometimes wanders and designers should ensure, when designing driven shoring where one sided concrete forming systems are anticipated, that driving tolerances are acceptable. Figure 3.24 is an example of driven soldier piling.

3.5.1.2 Drilled and Concreted Soldier Piling. The drilled and concreted soldier piling methods consist of drilling a hole of sufficient diameter to permit the introduction of a steel wide flange section with sufficient added space to overcome any variations from vertical (plumb) in the drilled hole. Once the hole is drilled, a steel wide flange section is introduced into the hole and hung to achieve verticality. The toe of the soldier pile (that portion of the pile which will always be below the base of the excavation) is backfilled either with structural concrete or with a lean sand grout such as CDF (Controlled Density Fill). The drilled shaft above the toe is usually backfilled with lean sand grout although applications for the use of fine gravel or sand as a backfill do exist. Typical soldier piles used in this application are 8 to 24 inch (200-610 mm) wide flange sections. Tables of typical soldier pile sections may be found in Chapter 18.2 and 18.3. Some soldier piles are fabricated from doubled channels or doubled wide flange sections (Figure 3.25). These doubled piles are discussed further under tieback connections in Chapter 4.4.

3.5.1.3 Churn Drilled Soldier Piles. In very difficult drilling conditions, such as wet bouldery gravels, soldier piles have been installed by churn drilling using 12 to 24 inch (305-610 mm) diameter pipe piles and then attaching the lagging to the pipes with welded clips (see Chapter 5). Churn drilling is used primarily in the well drilling industry. The process consists of taking a pipe and working a chopping bit inside the pipe. By adding water to the existing materials, and introducing slurrying agents as necessary, it is possible to pulverize the soils and rocks into a liquid mixture. The pipe is then tapped down into the mixture and the procedure is repeated until the pipe is advanced to its required depth. While the use of pipe is not particularly efficient from a bending moment capacity point of view, this can be a very effective way of placing soldier piles in a hostile environment.

3.5.1.4 Wet Set Soldier Piles. In recent years, soil cement mixing processes such as the Geojet or Soil Mixed Wall methods have been introduced in North America. Together these processes are referred to as Deep Mixed Method (DMM). The DMM consists of introducing a mixing wand into the soil which mixes the existing soils with cement and water to form soil cement. In the Geojet system, the mixing is by a combination of mechanical cutting and high pressure (2000 psi (13.8 Mpa)) grouting, while in the soil mixed wall method the mixing is by mechanical cutting and low pressure (150 psi (1 MPa)) grouting. After mixing a column of material, a soldier pile is introduced into the wet mixture and secured in place until cement

FIGURE 3.24 Driven soldier piles, Vancouver, WA. (*Courtesy of Condon-Johnson & Associates, Inc. Seattle, WA*)

FIGURE 3.25 Soldier pile and lagging utilizing a fabricated soldier pile made up of two wide flange section, Gorda, CA. (*Courtesy of ADSC—The International Association of Foundation Drilling, Dallas, TX*)

hydration begins. The soldier pile can usually be set into the wet soil cement under its own self-weight, but can be advanced, if necessary, with a small vibrator. The wet set method has been demonstrated to be a much faster process than conventional drilling and concreting. Piles set by this method have also been successfully extracted and reused, offering further economies.

3.5.2 Uses

Soldier pile and lagging is used in soils which exhibit sufficient arching potential to permit lagging (see Chapter 5). The soils must be above the static ground water table or have been dewatered. While soldier pile and lagging can tolerate seepage and leakage from seams, it is not designed to be used below the water table. In view of the fact that the lagging only extends to the base of the excavation, soldier pile and lagging is not applicable in soils which might exhibit basal instability (see Chapter 9.5).

In relatively stiff soils that have underlying slip failure planes, soldier piles can be designed to penetrate to sufficient depth to intersect and strengthen these slip planes.

3.6 SOIL NAILING

Soil nailing is a process that has been practiced in one form or another since Roman times. Very simply put, soil nailing consists of reinforcing the earth until a block of soil is created which is of sufficient size and strength to resist the overturning, sliding, and wracking forces applied to it by the lateral earth pressures. Figure 3.26 details a typical soil nail shoring wall. In principle, soil nailing is very similar to Mechanically Stabilized Earth (MSE), which is in use on many of our highway projects today providing retaining walls in fill situations.

Soil nailing merely extends this analysis to cut situations. In a typical soil nail designed wall, the soil is excavated in lifts and then near horizontal inclusions are placed into the soil at regular intervals to increase the shear strength of the soil and make it act as a block (see Figure 3.27). The reason that the inclusions are sub horizontal instead of horizontal is that some slight declination is necessary in order to keep the grout which is placed in the drilled hole, from pouring right back out. To complete the soil nailing process, the nailed face is covered with a shotcrete fascia. The process is repeated until the base of the excavation is reached. Figures 3.28 through 3.30 are photos of soil nail projects.

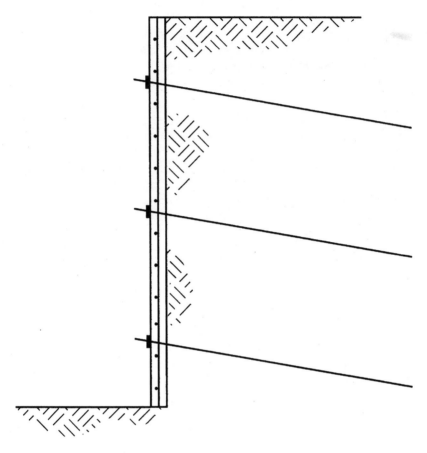

FIGURE 3.26 Typical section—soil nailing.

3.6.1 Uses and Limitations

Soil nailing is applicable in stiff soils such as overconsolidated clays, dense silts, medium to dense sands and gravels which show significant stand up time (see the discussion of standup time in Chapter 5), and cemented tills. It is not recommended in caving soils, nor softer cohesive soils. In order to improve the standup capability of the soil, nails are often drilled through a stabilizing berm prior to final cutting for shotcreting (see Figures 3.31 and 3.32). Soil nailing cannot address basal instability, but has been used very successfully to reinforce shallow slide planes (Chapter 6.5). When compared to soldier pile and lagging, soil nailing is almost always more economical.

TYPES OF SHORING SYSTEMS

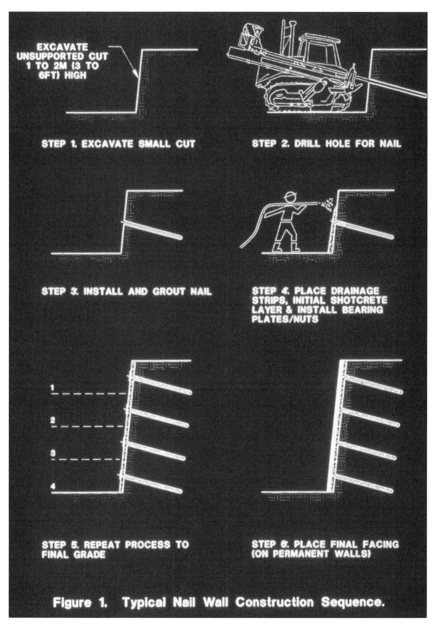

FIGURE 3.27 Soil nail sequence. *(Courtesy of Golder Associates, Inc. Redmond, WA)*

FIGURE 3.28 Soil nail wall under construction, Pocatello, ID. *(Courtesy of Condon-Johnson & Associates, Inc. Seattle, WA)*

FIGURE 3.29 Finished wall—top down soil nail wall, Boise, ID. (*Courtesy of Condon-Johnson & Associates, Inc. Seattle, WA*)

FIGURE 3.30 Finished top down soil nail wall, Boise, ID. Note the sensitive utility adjacent to cut. (*Courtesy of Condon-Johnson & Associates, Inc. Seattle, WA*)

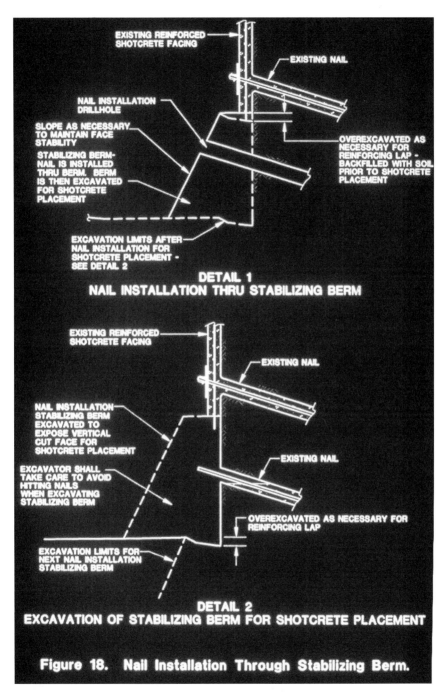

FIGURE 3.31 Drilling sequence through a berm—schematic. *(Courtesy of Golder Associates, Inc. Redmond, WA)*

FIGURE 3.32 Drilling through a berm, Redmond, WA. *(Courtesy of Condon-Johnson & Associates, Inc. Seattle, WA)*

3.7 SECANT PILES

Secant piles are drilled shafts which are interlocked (see Figure 3.33) to form a continuous wall. The wall is constructed by drilling alternate shafts and then back stepping to drill the intervening shafts in order to interlock the two adjacent shafts. Figure 3.34 demonstrates the type of interlock which can be obtained. Every second shaft is reinforced, usually with a wide flanged steel section or alternatively with a reinforcing steel cage. The reinforced shafts are called "primaries" or "king" piles. The intervening piles are not reinforced and are called "intermediates" or "secondaries."

The drilling sequence usually calls for the intermediates to be drilled first. This is done so that the reinforcing of the primary piles will not be compromised by subsequent drilling. The concrete used for the secondary piles is lean concrete. Lean concrete is used so it will remain soft enough for the drilling and interlocking of the primaries. The primaries are drilled after the secondaries have gained sufficient strength to permit the adjacent drilling. The primary piles can be poured with either lean concrete or structural if the reinforcing is by wide flanged beam. If the reinforcement of the primary is with a reinforcing steel cage, the primary will always be poured with structural concrete. In cases where the secant wall is formed by DMM, the primaries are always reinforced with wide flanged beams. Figures 3.35 through 3.37 are examples of completed secant pile walls.

When the purpose of the secant wall is to retain water or saturated soils, the lean concrete mix should have a compressive strength of about 500 psi (3.5 MPa). If the wall is retaining unsaturated soils and is not required to retain water, a strength of about 150 psi (1 MPa) may be allowed. Figure 3.37 is a photo of a secant wall built utilizing lean mix of approximately150 psi (1 MPa). Note that the contractor was able to shave the face of the secants in order to present a flatter surface to form and pour concrete against.

In the case of water bearing soils, the secondary piles are extended to the same depth as the primary piles in order to create a cutoff wall (see Figure 3.38). If the application of the wall is for shoring soils where water movement is not a problem, then the secondary piles are normally terminated about one foot below the level of the base of the excavation (see Figure 3.39).

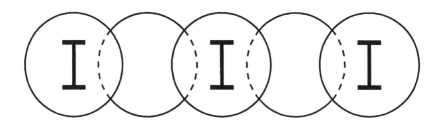

FIGURE 3.33 Secant wall plan—schematic.

FIGURE 3.34 Secant wall with good interlock, Toronto, Ontario. (*Courtesy of Deep Foundations Contractors. Thornhill, Ont.*)

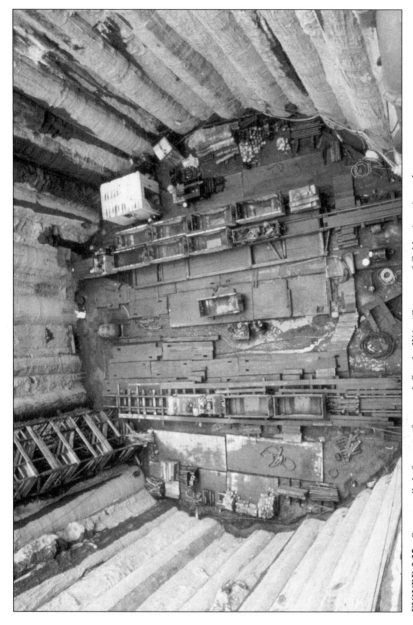

FIGURE 3.35 Secant wall-shaft access for tunnel, Seattle, WA. *(Courtesy of Golder Associates, Inc. Redmond, WA)*

FIGURE 3.36 Secant wall against weak rubble foundation, Toronto, Ont. (*Courtesy of Deep Foundations Contractors. Thornhill, Ont.*)

FIGURE 3.37 Secant wall of lean mix shaved to present flat face for subsequent cast-in place concrete wall, Toronto, Ont. *(Courtesy of Deep Foundations Contractors. Thornhill, Ont.)*

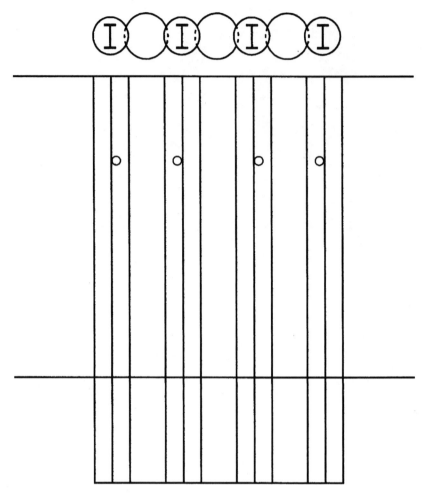

FIGURE 3.38 Secant wall designed to cutoff water flow below the excavated depth.

3.7.1 Uses

Secant walls may be used in any of the situations which are suitable for sheet piling except retaining open water. According to the ASCE GSP # 74, Guidelines of Engineering Practice for Braced and Tied-Back Excavations, secant walls are suitable as a water cutoff to a depth of about 40 feet (12.2 m). Beyond this depth, problems are encountered in maintaining shaft interlock because of drilling tolerances. Some of these problems can be overcome by tightening up the spacing of the secant piles and increasing the overlap.

TYPES OF SHORING SYSTEMS 65

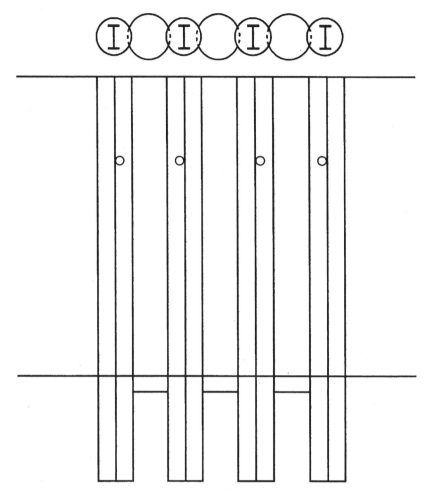

FIGURE 3.39 Secant wall designed for earth retention where no water cutoff is necessary.

When compared to soldier pile and lagging walls, secant walls have been found to be particularly effective in situations where minor loss of soil during lagging operations might be detrimental to adjacent footings or sensitive utilities.

Because secant piles can be reinforced with wide flanged sections, they often can be designed with greater moment resistance than sheet piles. This, coupled with the fact that they are drilled and not driven, gives the advantage to secant walls in situations where vibrations might be detrimental, where walls must be installed very close to adjacent buildings or where a more rigid cutoff is required to ensure basal stability (Chapter 9.5).

FIGURE 3.40 Tangent pile wall for slide prevention, Seattle, WA. *(Courtesy of Condon-Johnson & Associates, Inc. Seattle, WA)*

3.7.2 Tangent Piles

Looking very similar to secant pile walls, tangent pile walls are constructed with the edges of the drilled shafts (their tangents) just touching each other (see Figure 3.40). This type of wall will not function as a water barrier, but is quite efficient in those situations where the primary reason for choosing an alternative to soldier pile and lagging is to ensure that soil loss during excavation does not occur.

3.8 CYLINDER PILE WALLS

Cylinder pile walls are really cantilevered tangent piles, "super sized." These walls are used primarily in highway side hill cut situations in order to ensure that sliding of the undercut hillside does not occur. Cylinder piles are drilled in diameters of 6-10 feet (1.8-3.0 m) and reinforced either with heavy, specially fabricated girder sections, or heavy reinforcing cages (see Figure 3.41).

The toe of the cylinder pile is designed to restrain not only the active pressures from the excavated face, but also to intercept and strengthen any slide planes which may affect the hillside stability (Figure 3.42).

Once the cylinder piles are installed, the excavation can be performed. A cap beam and fascia wall is attached to the exposed portion of the wall (see Figure 3.43). This solution has been used to cantilever walls on highways of up to 25 feet (8 m) in height.

TYPES OF SHORING SYSTEMS 67

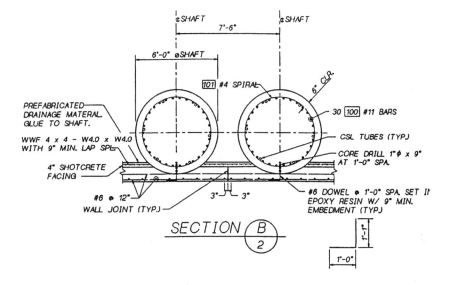

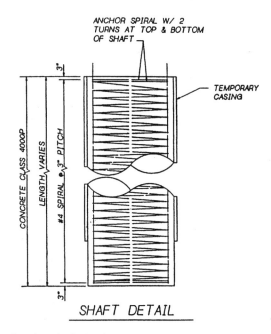

FIGURE 3.41 Typical plan and section of cylinder pile wall, Auburn, WA. *(Courtesy of Washington State Department of Transportation)*

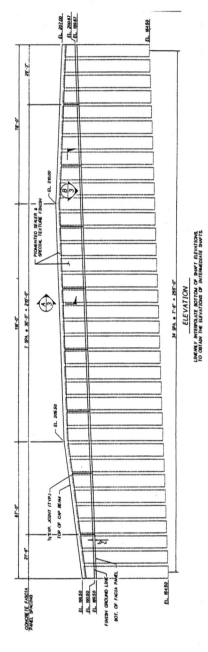

FIGURE 3.42 Elevation—cylinder pile wall construction, Auburn, WA. *(Courtesy of Washington State Department of Transportation)*

FIGURE 3.43 Completed cylinder pile wall with shotcrete fascia, Auburn, WA. *(Courtesy of Condon-Johnson & Associates, Inc. Seattle, WA)*

3.9 SLURRY WALLS

Slurry Walls are cast-in-place concrete walls (see Figure 3.44) constructed prior to the excavation of the site. These walls almost always serve the dual purpose of shoring the site during excavation and acting as the permanent wall once the structure is complete (see Chapter 6.2).

Slurry walls are constructed by excavating primary and secondary slots or trenches (Figure 3.45) of approximately 20-30 feet (6.1-9.1 m) in length to the ultimate depth desired. The trench is stabilized with the introduction of mineral or polymer slurry. The trench is usually excavated with specially developed digging buckets or clams. Extremely difficult conditions, including rock and nested boulders, may be excavated using specially designed tools with rotating cutter heads called "hydrofraizes."

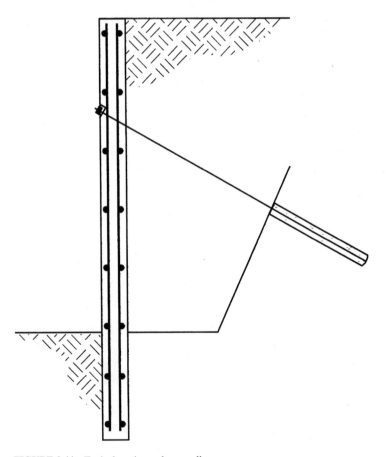

FIGURE 3.44 Typical section—slurry wall.

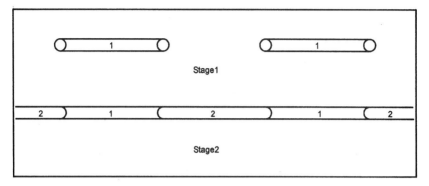

FIGURE 3.45 Typical slurry wall indicating excavation sequence.

In order to guide the digging tools into digging the trench guide walls (Figure 3.46) are first constructed of cast-in-place concrete. Once the trench is dug to depth, a reinforcing steel cage is lowered into the trench and concrete is tremied into the slurry to cast the wall. As the concrete displaces the slurry, the slurry is pumped off into tanks or into adjacent trench excavations.

In order to pour a wall with a regular end configuration, end stops are placed in the primary slots after completion of excavation and prior to introduction of the reinforcing steel. Originally end stops were made of heavy steel pipe. Recent developments have utilized fabricated sections which can introduce waterstop elements into the joints (Figure 3.47). Concrete is tremied against the end stops and the end stops are pulled after the concrete has taken its initial set. The resultant shape at the end of the primary panel forms a female joint (Figure 3.45) for the secondary panel to be poured into, which creates a waterproof joint.

Figure 3.48 is a photo of a finished slurry wall. Where aesthetic finishes are desired, a cast-in-place or precast fascia will be used to cover the exposed face.

3.9.1 Uses

Slurry Walls are used in situations which are similar to those of secant walls. They are recommended as a waterproof solution to depths of 100 feet (30.5 m) by the ASCE GSP # 74 and have been used to depths of 400 feet (122 m) as dam cutoff walls. In view of the length of trench open at one time, slurry walls are not recommended adjacent to shallow spread footings. Because slurry walls are quite costly, permanent wall construction utilizing slurry wall techniques is usually restricted to waterbearing soils which are very difficult to drill.

FIGURE 3.46 Poured guide walls frame and excavated slurry wall panel—end stop in place. *(Courtesy of Hans Leffer GmbH. Saarbrucken, Germany)*

3.10 MICROPILE WALLS

When engineers are faced with construction of a permanent retaining wall which is too high to cantilever and not blessed with sufficient right of way for either soil nailing or tiebacks, micropiles can be used to support the wall. Micropiles are small diameter, high capacity, drilled piles that derive their capacity through pressure grouting techniques.

Micropiles are installed in an A frame type of arrangement (see Figure 3.49). Utilizing duplex drilling methods, one line of micropiles is battered and the other installed vertical (see Figure 3.50). The piles are tied together in a cap which forms a moment connection (see Figure 3.51).

FIGURE 3.47 Close up of end stop with water stop in place. *(Courtesy of Hans Leffer GmbH. Saarbrucken, Germany)*

The wall is excavated in stages. Studs are welded to the face of the vertical micropiles and reinforcing steel for wall fascia construction placed (see Figure 3.52). Shotcrete is then applied to complete the retaining wall (see Figure 3.53). The shotcrete can be installed as the exposed face by finishing it, but in this case a cast-in-place fascia was added to incorporate a textured finish.

3.11 UNDERPINNING

Underpinning, when used in concert with shoring, is performed to support adjacent structures while excavation is carried out directly beside the building footings. If the adjacent structure has a footing perched at a depth which is shallower than the proposed excavation, the excavation could compromise the bearing capacity of the adjacent footing unless specialized treatment of that footing is undertaken. This treatment is called underpinning, and can be performed by a variety of methods.

FIGURE 3.48 Slurry wall after construction of cap beam but prior to placement of fascia wall, Mercer Island, WA. (*Courtesy of Golder Associates, Inc. Redmond, WA*)

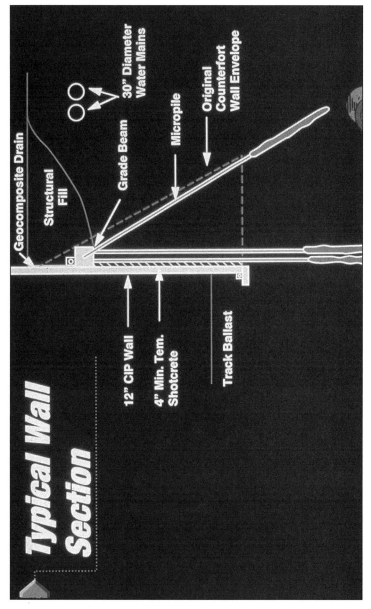

FIGURE 3.49 Typical section, A-frame micropile retaining wall. (*Courtesy of Golder Associates, Inc. Redmond, WA*)

FIGURE 3.50 Drilling micropiles for retaining wall, Portland, OR. (*Courtesy of Golder Associates, Inc. Redmond, WA*)

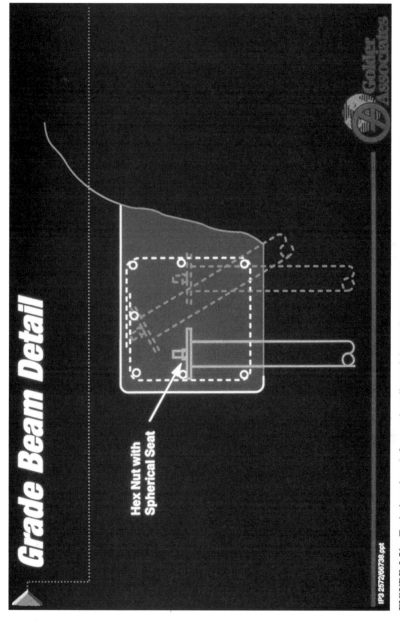

FIGURE 3.51 Typical section, A-frame micropile retaining wall cap beam. (*Courtesy of Golder Associates, Inc. Redmond, WA*)

FIGURE 3.52 Fascia construction attaching shotcrete wall to micropiles with studded connection, Portland, OR. *(Courtesy of Golder Associates, Inc. Redmond, WA)*

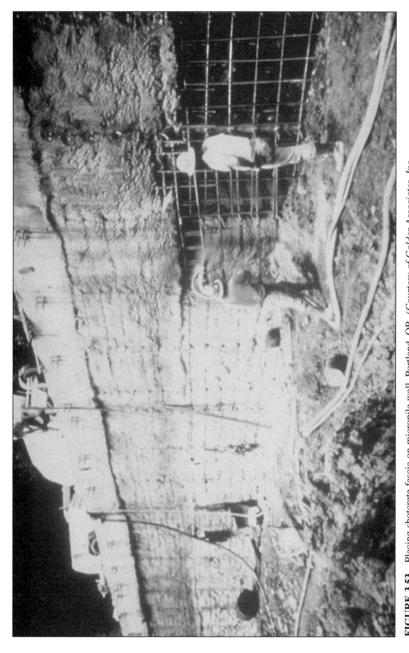

FIGURE 3.53 Placing shotcrete fascia on micropile wall, Portland, OR. *(Courtesy of Golder Associates, Inc. Redmond, WA)*

3.11.1 Panel Underpinning

Underpinning depths of less than 10 feet (3.0 m) can be handled by panel underpinning. In this method, panels are excavated, and formed, and concrete is poured to extend the footing of the shallow foundation to the base of the new excavation. The poured panel is stopped 2 inches (50 mm) below the footing being underpinned and the resultant gap is dry packed with cement grout to ensure tight contact. Panels are excavated in an alternating fashion so that at all times the footing is being supported. Panels can be excavated before (see Figure 3.54) or after mass excavation (see Figure 3.55).

This method requires that the soil being excavated exhibit good standup time. This is an absolute requirement so that the sides of the panel can be true and that ground is not lost from under the slab on grade behind the footing being underpinned. A completed panel underpinning scheme is shown in Figure 3.56.

3.11.2 Underpinning Pits

Underpinning piers can be constructed under adjacent building footings by digging pits. These pits are shored as they are excavated, in a fashion similar to hand excavated caissons. Once a pit excavation is completed, it is filled with structural concrete and the gap between the top of the pit pour and the underside of the footing is filled with dry packed grout.

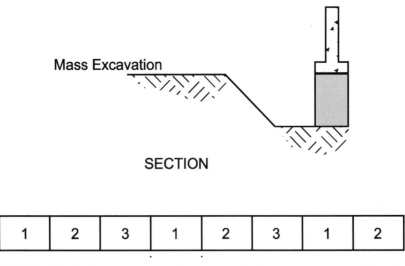

FIGURE 3.54 Typical panel underpinning—constructed prior to mass excavation. Note the sequence of the panel construction.

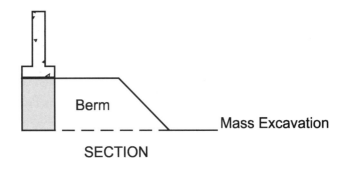

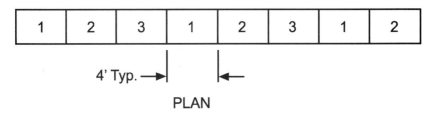

FIGURE 3.55 Typical panel underpinning—constructed after mass excavation.

When the pit piers are in place, excavation can proceed. Lagging boards are placed between the adjacent underpinning piers to ensure that ground is not lost from under the building floor slab (see Figure 3.57).

Soil conditions necessary for satisfactory completion of this type of underpinning are dry or dewatered soils suitable for pit excavation where pit shoring can be installed at least one foot (300 mm) at a time without ground loss.

3.11.3 Slant Piles (Figure 3.58)

Slant pile underpinning consists of drilled soldier piles which are excavated under the adjacent footing by drilling a shaft adjacent to the footing and angling the shaft so that its base will be directly below the footing being underpinned (see Figure 3.59). A soldier pile is then installed vertically in the shaft. Some hand excavation or reaming is required to advance the shaft under the footing to permit placement of the soldier pile in its vertical position beneath the footing.

The load of the footing is transferred to the soldier pile through a welded plate on the top of the pile which is dry packed to the underside of the footing (see Figure 3.60). Once the pile is drypacked, the excavation adjacent to the underpinned footing may commence with lagging and tiebacks being installed as required (see Figure 3.61).

FIGURE 3.56 Panel underpinning, Toronto, Ont. *(Courtesy of Deep Foundations Contractors. Thornhill, Ont.)*

FIGURE 3.57 Hand excavated underpinning piers, Denver, CO. *(Courtesy of Schnabel Foundation Co., Inc. Houston, TX)*

84 EARTH RETENTION SYSTEMS

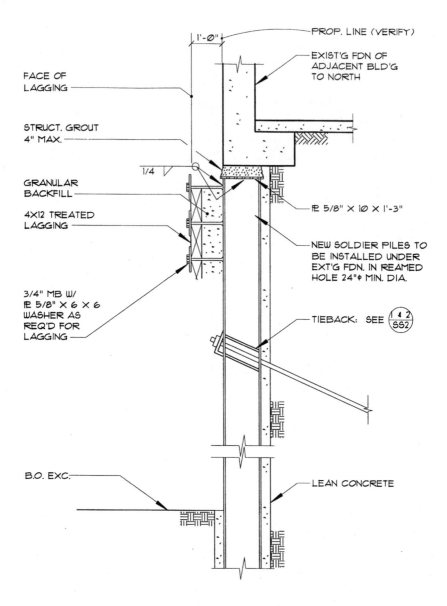

FIGURE 3.58 Soldier pile underpinning utilizing slant drilling techniques—typical section. *(Courtesy of CT Engineering, Seattle, WA)*

FIGURE 3.59 Drilling slant pile underpinning, Seattle, WA. *(Courtesy of Condon-Johnson & Associates, Inc. Seattle, WA)*

Slant pile underpinning has been successfully installed to depths of 80 (24 m) feet, but must be drilled in materials which can stand without casing (see Figure 3.62). It simply is not possible to case a shaft drilled under an adjacent footing. If necessary, dewatering must be undertaken to ensure that hole instability does not cause loss of ground under the adjacent building.

3.11.4 Soldier Piles with Corbels

In situations where it is not necessary to locate the underpinning directly under the adjacent building, or not possible because of casing requirements, it is possible to drill or drive a soldier pile adjacent to the footing and underpin the structure by attaching a corbel to the pile and dry packing it under the footing (see Figure 3.63).

FIGURE 3.60 Slant drilled underpinning pile with dry packing to building footing, Seattle, WA. *(Courtesy of Condon-Johnson & Associates, Inc. Seattle, WA)*

FIGURE 3.61 Construction of shoring once slant underpinning piles are installed, Seattle, WA. *(Courtesy of Condon-Johnson & Associates, Inc. Seattle, WA)*

FIGURE 3.62 An 80 foot underpinning by slant pile techniques, Seattle, WA. *(Courtesy of City Transfer, Inc. Kent, WA)*

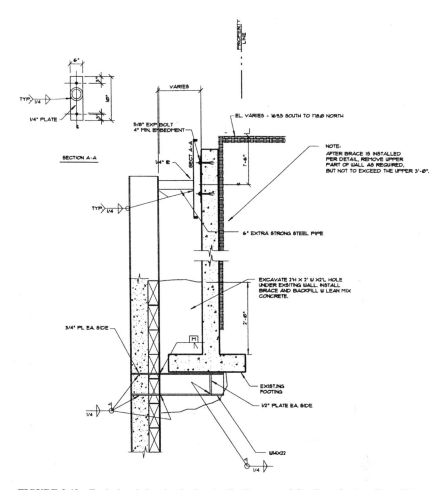

FIGURE 3.63 Typical corbel underpinning detail. *(Courtesy of City Transfer, Inc. Kent, WA)*

An alternate to this method can be constructed utilizing micropiles (see Figure 3.64). On this particular project, micropiles were installed and then capped with concrete pile caps attached to the adjacent building footing with epoxy dowels. The excavation was then progressed utilizing soil nailing. The completed underpinning scheme is shown in Figure 3.65.

3.11.5 Jacked Piles

Underpinning can be effected with the use of jacked piles. Pipe piles (open or closed ended) or H piles can be jacked into location below the footing intended to be underpinned. In the case of a wall footing, a small pit is excavated below

FIGURE 3.64 Corbel construction by epoxying dowels into footing being underpinned, Salt Lake City, UT. Note micropile installed to provide underpinning support. (*Courtesy of Condon-Johnson & Associates, Inc. Seattle, WA*)

FIGURE 3.65 Square protrusions at base of existing building footings are completed corbels which incorporate micropiles, Salt Lake City, UT. *(Courtesy of Condon-Johnson & Associates, Inc. Seattle, WA)*

the footing. By jacking against the dead weight of the structure being underpinned, the piles are advanced to the required depth (see Figure 3.66). Piles are spliced by welding or by the use of manufactured pile splicers.

Once the pile is jacked to the desired depth, an attachment is made between the pile and footing and the jack load is removed. In this way the load is transferred from the footing to the pile without permitting settlement. Care must be taken when advancing the pile that the jacking loads do not exceed the dead weight of the structure. If this occurs, uplift forces will be exerted on the structure with possible damage resulting.

The engineer designing a jacked piling system must ensure that the capacity of the pile is developed entirely below the anticipated depth for the proposed excavation. For this reason, the capacity of the pile cannot be inferred directly from jack loads as the friction in the excavation zone must be discounted.

Because pile friction in the excavation zone must be discounted and the jacking forces kept below the dead load of the structure, it is often necessary to utilize pile groups to provide sufficient capacity. Group effects must also be calculated when designing this type of underpinning.

FIGURE 3.66 Jacked underpinning piles, Kitchener, Ont. *(Courtesy of Deep Foundations Contractors. Thornhill, Ont.)*

FIGURE 3.66 *(continued)* Close up of jacked underpinning piles, Kitchener, Ont. *(Courtesy of Deep Foundations Contractors. Thornhill, Ont.)*

CHAPTER 4
LATERAL SUPPORT

In any situation involving a retaining wall or shoring structure, lateral loads exist. The applied loads may be the result of earth pressures, seismic loads, surcharge loads, or hydrostatic pressures. But, whatever their source, they are constantly trying to push the wall over and must be restrained. The restraint can be developed from inside the excavation or outside. Commonly used methods are few in number but many in their variations.

Rakers are sloping compression units that derive their capacity inside the excavation (see Figure 4.1). They are attached to the wall and braced against either the structure being constructed, or a footing specifically cast for the purpose of resisting the raker forces. Since rakers are sloping elements, they impart not only a lateral force to the wall to counteract the applied load, but also an uplift force. This uplift force is counteracted by friction, either above the base of the excavation or below the excavation if the wall has a toe element (that portion of the wall which extends below the base of the excavation).

Struts are another bracing type which function from within the excavation. Struts are horizontal compression units which attach to the wall normal to the imposed lateral load (see Figure 4.2). Struts are braced against either an existing structure or another portion of the shoring system. Because the strut is applying a horizontal force at right angles to the wall, uplift loads are not a concern.

Deadman anchors are tension elements which restrain the applied load from outside the excavation (see Figure 4.3). The deadman tendon attaches to a buried anchorage and applies a horizontal restraining force.

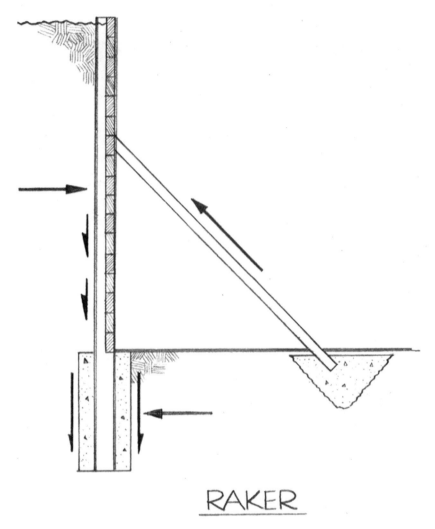

FIGURE 4.1 Typical section—soldier pile and raker construction.

Tiebacks are tension units similar to deadman anchors except that the tieback is constructed with a slight downward slope (see Figure 4.4). Attached at right angles to the wall in the horizontal plane, tiebacks derive their capacity from friction between the tieback and the soil or rock in which it is embedded. Since tiebacks are installed at a downward dipping angle, they also impart a downward force to the wall which must be counteracted either by friction behind the wall facing, or through a combination of friction and end bearing in the toe of the principal wall element.

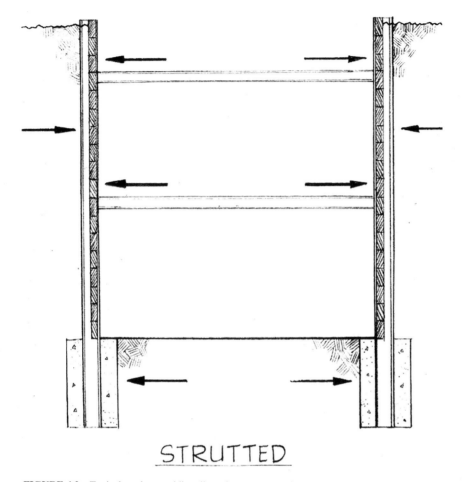

FIGURE 4.2 Typical section—soldier pile and strut construction.

Soil nailing restrains the applied lateral pressures by the mobilization of gravity forces (see Figure 4.5). The process creates a soil mass that is sufficiently rigid and will act as a unit to resist the lateral loads applied to it. The weight of the block when taken as a moment about the leading edge of block (Pt 0) resists overturning. The base of the block is of sufficient area that it will resist sliding through friction on the base. Because of its reinforcement, the block of soil has sufficient shear strength to resist wracking. While this method derives its capacity from outside the excavation, it is the only method presented which does not have toe elements.

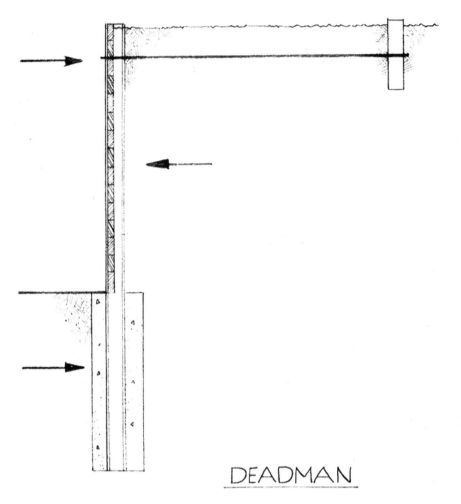

FIGURE 4.3 Typical section—soldier pile and deadman construction.

Cantilever shoring derives its capacity to resist lateral loads through embedment. The toe element of the wall is embedded to a sufficient depth that a point of rotation occurs. Forces on either side of this point of rotation form a moment couple which resists overturning (see Figure 4.6).

A number of structural variations exist to effectively utilize these restraint methods but these six methods are evident in any number of combinations in order to create capacity to resist lateral load. Development of loads and forces will be dealt with in Chapters 9 and 11.

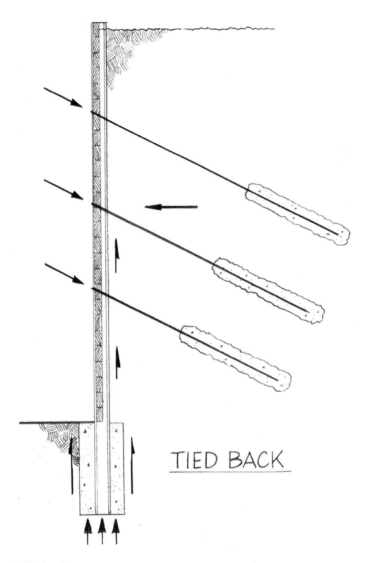

FIGURE 4.4 Typical section—soldier pile and tieback construction.

4.1 RAKERS

Rakers are compression members which almost always are steel although, in some shallow excavations, timber rakers are used. Designed as columns against buckling, square wide flange beams or pipes are most often utilized (see Figure 4.7). Rakers, which are installed in footings designed exclusively for that purpose, are usually installed at 45 degrees to the horizontal.

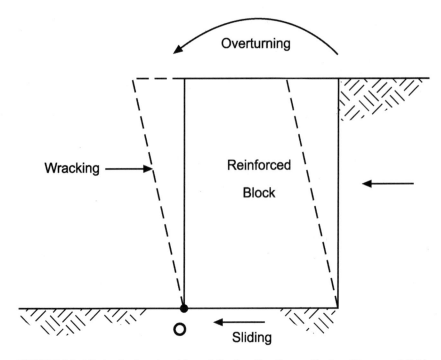

FIGURE 4.5 Block of soil analyzed for stability in soil nailing application. *(Courtesy of Golder Associates, Inc. Redmond, WA)*

In order to install the raker, it is necessary to dig to the base of the excavation. To do this and still retain the wall in its installed position, not all the excavation can be performed. Figure 4.7 details a raker with its berm which must remain in place until the raker is installed. This berm must provide the passive resistance required to allow the wall to function in cantilever. In spite of this berm, movements of the wall into the excavation will occur. The amount of movement is inversely proportional to the size of berm used to restrain the wall.

If the raker is braced against a portion of the structure being constructed, the angle of declination of the raker is usually on the order of 35 degrees. This allows the contractor to construct the base slab of the structure and use some room to form the slab edge without impinging on the berm. Rakers are either welded to the wall (Figure 4.8) or fitted into weldments. Beams are cut to fit the element wall (soldier pile, sheet pile, or waler). If pipe rakers are used, (see Figure 4.9) a plate is usually installed in the end of the pipe to be attached to the wall in the line of the axis of the pipe. This plate is welded into the pipe and then cut to fit the wall element for welding to the wall (see Figure 4.10).

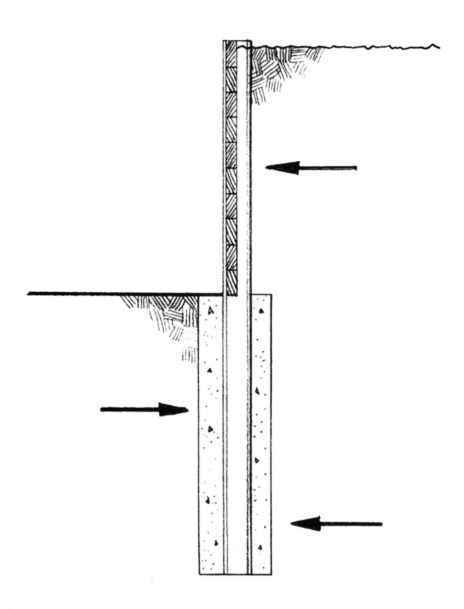

FIGURE 4.6 Typical section—cantilever soldier pile construction.

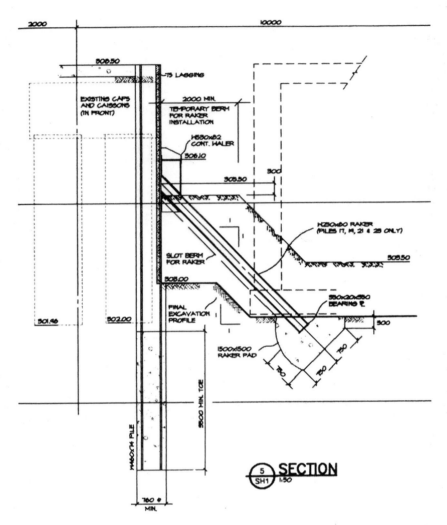

FIGURE 4.7 Typical raker section, Toronto, Ont. *(Courtesy of Isherwood Associates. Oakville, Ont.)*

Raker footings are usually unreinforced mass concrete which are narrow and deep. They can also be constructed as drilled shafts with a steel element cast in them to allow attachment of the raker. Figure 4.11 exhibits drilled raker footings. Preloading of rakers is often undertaken in order to restrict the movement of walls being braced by rakers. Large movements often occur to walls while the rakers are being installed, some of which can be recovered by jacking. The preloading of rakers is performed by jacking and welding which is labor intensive and adds significant cost to the shoring system. Once the raker is installed, and preloaded if specified, the berm can be removed.

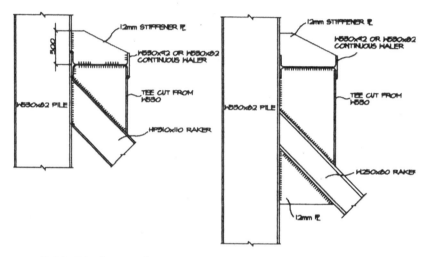

FIGURE 4.8   Typical raker direct connection to wall, Toronto, Ont. *(Courtesy of Isherwood Associates. Oakville, Ont.)*

Rakers must remain in place to provide lateral support to the wall until such time as the structure being constructed within the excavation can accept that load. Because of this, rakers must be left in place while construction progresses and the structure must be built around the raker. This involves blocking out formwork to permit passage of the rakers through floors and walls. Once the structure is complete, the rakers are cut out, often in pieces, and the area left is patched.

4.1.1 Rakers and Walers

Because rakers interfere with formwork and can be difficult to excavate around, walers are often integrated with rakers in order to minimize the number of rakers installed. Walers (also called wales) are wide flange steel beams which are attached horizontally to the wall. The walers are designed as bending elements and

FIGURE 4.9 Pipe raker, Seattle, WA. *(Courtesy of Condon-Johnson & Associates, Inc. Seattle, WA)*

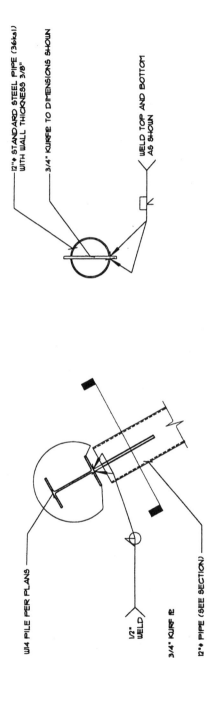

FIGURE 4.10 Typical connection detail—pipe to beam. *(Courtesy of Schnabel Foundation Co., Inc. Houston, TX)*

FIGURE 4.11 Rakers supported on drilled footings, Seattle, WA. (*Courtesy of Condon-Johnson & Associates, Inc. Seattle, WA*)

distribute the horizontal forces from the raker to the wall. Walers can be attached directly to the wall (see Figures 4.12, 4.13). This is the simplest and most economic attachment method. It is used in cases where the waler will not interfere with the structure being built. Often waler systems are designed to be located just above a floor level so that once the floor is poured, the waler can be removed.

When walers are directly in contact with the wall, the wall/waler connection, when not complicated by rakers or struts, is in simple axial compression. As such, it is not necessary to weld this connection. Often the gap between the waler and wall is filled with wooden or steel wedges.

If it is not possible to locate the waler so that it does not interfere with the proposed structure, the waler can be installed inside the structure. The waler is attached to the wall by the use of stubs (see Figures 4.14 and 4.15). The structure can then be constructed by boxing out around the stubs and the waler removed from inside the structure at the appropriate time.

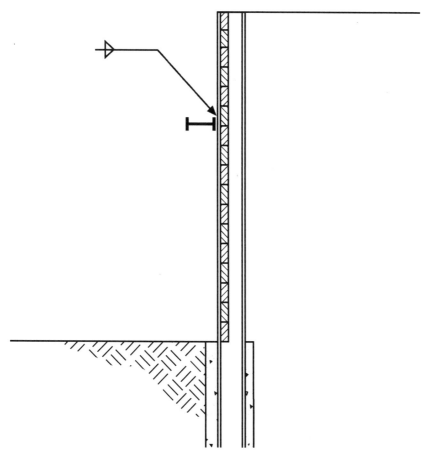

FIGURE 4.12 Typical section—waler mounted directly to wall.

FIGURE 4.13 Waler mounted directly to Geojet secant wall, San Francisco, CA. (*Courtesy of Condon-Johnson & Associates, Inc. Oakland, CA*)

LATERAL SUPPORT

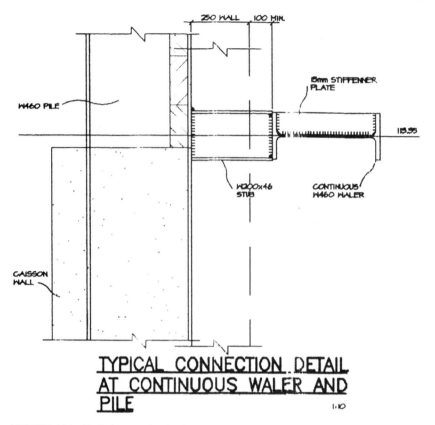

FIGURE 4.14 Typical connection detail—waler to wall by the use of stubs to offset waler from wall to permit wall construction. *(Courtesy of Isherwood Associates. Oakville, Ont.)*

4.1.1.1 Raker to Waler. If the waler is attached directly to the wall, the raker can be attached to the outer flange of the waler. Because of the direction of the load applied, the waler will have a tendency to roll upwards. This is overcome with the use of a roll chock (see Figures 4.16 through 4.18).

4.1.1.2 Raker to Waler and Pile. If the waler is attached directly to the wall, the raker can be attached to the inner flange of the waler and pile simultaneously (see Figures 4.19, 4.20). This method relieves the designer of the necessity to deal with the torsional loading of the waler.

FIGURE 4.15 Waler to wall on stubs, Toronto, Ont. *(Courtesy of Isherwood Associates, Oakville, Ont.)*

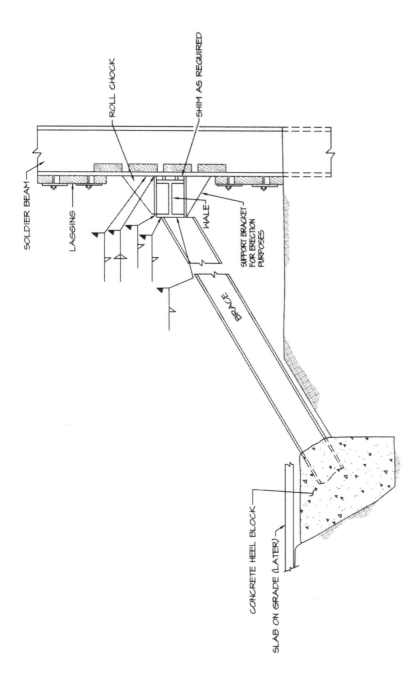

FIGURE 4.16 Typical connection of raker to waler. Note the use of a roll chock which prevents the waler from rolling under raker load. (*Courtesy of Schnabel Foundation Co., Inc. Houston, TX*)

FIGURE 4.17 Raker connection direct to waler-roll chocks not yet installed. Note raker installed through berm, Houston, TX. (*Courtesy of Schnabel Foundation Co., Inc. Houston, TX*)

FIGURE 4.18 Close up of roll chock. *(Courtesy of Schnabel Foundation Co., Inc. Houston, TX)*

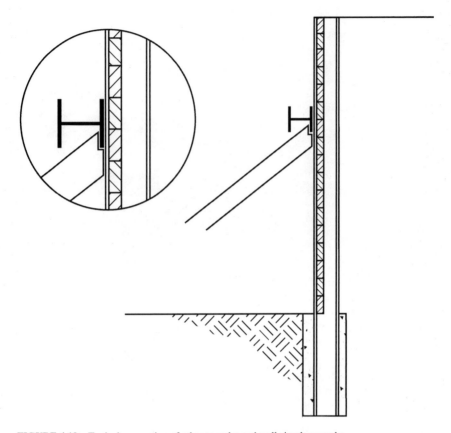

FIGURE 4.19 Typical connection of raker to waler and wall simultaneously.

4.1.1.3 Raker to Waler then Pile. If the waler is set out on stubs, the raker can be attached directly to the wall and the waler attached to the raker by use of a lookout or supporting stub (see Figures 4.21 and 4.22.) The lookout forms a convenient erection template. Once the rakers are installed and the lookouts attached to them, the waler can be laid out on the lookouts and manipulated by hand for final fit-up. A plate attachment is then welded to the waler and pile in the axis of the web of the raker. This type of connection is not nearly as simple if the raker is a pipe section and so is not often used with pipe rakers.

FIGURE 4.20 Raker to waler and wall, Toronto, Ont. Note that raker contacts waler and wall simultaneously. *(Courtesy of Isherwood Associates. Oakville, Ont.)*

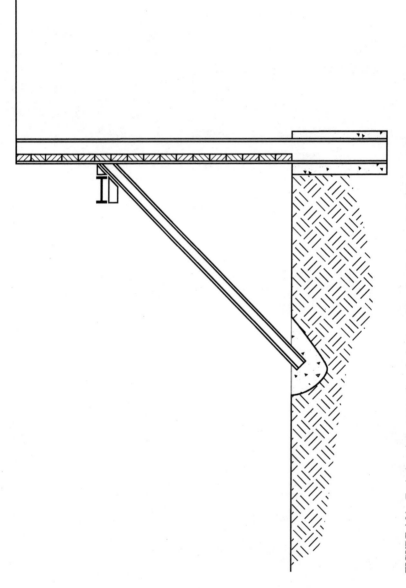

FIGURE 4.21 Typical section—connection of raker to waler and wall when waler is offset on stubs.

FIGURE 4.22 Raker and waler on stubs, Toronto, Ont. *(Courtesy of Deep Foundations Contractors. Thornhill, Ont.)*

4.2 STRUTS

Struts are usually associated with trench excavation where the depth or length of time required for the trench to be open requires substantial shoring. These types of excavations are called cut and cover. Some shoring walls may be strutted against existing structures, but the vast majority of struts are braced against the opposite side of the excavation. Since the walls of the cut and cover are parallel, the struts can be installed as axial compression elements with little or no design eccentricity. These struts can be attached to the piles by welding or can be wedged. If wedging is used, a careful analysis of the types of end treatments must be made. Since excavation must be carried out under the struts, and hoisting must pass through the struts, accidental striking of a strut cannot be permitted to dislodge the strut.

Lightly loaded struts can be of timber or telescoping pipe. More substantial struts are usually wide flange column sections. As the strut loads increase, and the width of the trench expands, large diameter pipes (24"-36" (610-915 mm)) are used. In some cases, a row of supports must be installed in the center of the trench to support the struts and limit their unsupported length. In an extreme case, the author worked on a project which was 50 feet (15.2 m) deep and 150 feet (46 m) wide. The excavation was braced with only one row of struts which were very heavy trusses constructed from wide flange beams.

As previously mentioned, because struts must permit excavation under them and hoisting through them, it is normal to find walers spanning the length of the wall. The wales are periodically braced across the trench by strutting. This arrangement minimizes the number of struts crossing the trench.

4.2.1 Struts and Walers

4.2.1.1 Strut Under Waler. In this configuration, struts are installed across the trench at predetermined intervals from soldier pile to soldier pile or sheet pile to sheet pile. Walers are then laid over the struts and attached as indicated in Figures 4.23 and 4.24. The struts form a template for the walers. The waler can then be blocked or wedged to restrain the remainder of the wall in the same manner as discussed in Section 4.1.1.

4.2.1.2 Strut to Waler. In this arrangement, struts are attached directly to the waler and transmit their force through the waler web. This detail calls for very true alignment of the strut to the waler web. Regardless of the accuracy of the alignment, the wale will have a tendency to roll either up or down and this must be restrained. See Figure 4.25 for a drawing of a welded wale strut connection with anti-roll chock. Figures 4.26 and 4.27 are photos of the same type of connection.

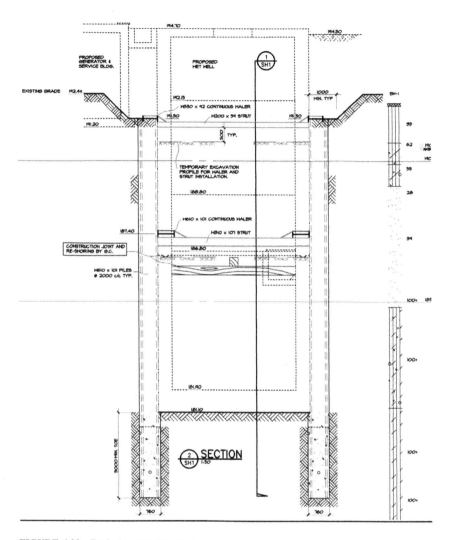

FIGURE 4.23 Typical connection detail when strut is mounted under waler. *(Courtesy of Isherwood Associates. Oakville, Ont.)*

FIGURE 4.24 Strut under waler, Toronto, Ont. *(Courtesy of Isherwood Associates. Oakville, Ont.)*

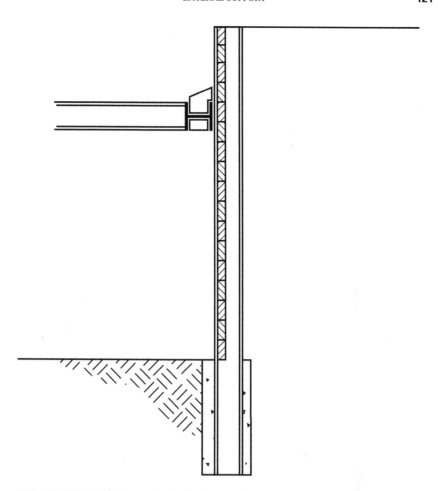

FIGURE 4.25 Typical connection detail when strut is mounted against waler. Note the roll chock. *(Courtesy of Golder Associates, Inc. Redmond, WA)*

FIGURE 4.26 Strutted excavation with strut mounted against waler, Niagara Falls, NY.

FIGURE 4.27 Strutted excavation with strut mounted against waler, San Francisco, CA. *(Courtesy of Condon-Johnson & Associates, Inc. Seattle, WA)*

The strut to waler connection lends itself to the assembly of bracing sets (see Figure 4.28). The sets consist of parallel walers with welded struts forming bracing rectangles. These rectangles are lowered into the trench between parallel rows of sheet piling or soldier piles. In fact, the frames can be laid on the ground first and used as a driving template for the sheet piles. The frames are manufactured slightly narrower than the planned trench width. Once the brace is set in place, it is wedged tightly against the shoring walls and suspended on chains to prevent it from slipping. Excavation can then progress. When the utility installation is complete, the trench is backfilled. Once the backfill reaches the height of the bracing set, the wedges are knocked free and the bracing rectangle removed for reuse.

An economic form of attachment of waler to strut is found in Figure 4.29. In this case the strut is installed in the trench between parallel walers. The strut is designed to be slightly (say 3 inches (75 mm)) shorter than the length required and the gap is filled with grout. This also permits the easy removal of the strut when it is no longer required.

4.3 CORNER BRACES

Where shoring walls face inwards at 90 degrees to each other and intersect forming a corner, an opportunity is presented for corner bracing. Corner braces provide lateral restraint to each wall in a manner similar to struts with one important difference. Corner braces also impart a horizontal lateral force to the wall which must be dealt with.

Corner braces can be quite small when used to brace areas of the wall close to a corner. Figure 4.30 details corner braces of less than 20 feet (6.1 m) in length which are welded directly to the soldier pile. Because the corner braces impart a lateral force into the wall, the force must be translated down the wall to dissipate the load through wall/soil friction or to the next corner where it can be resisted by the corner. In this instance, the waler is actually a square tube section mounted within the flange of the soldier piles (Figure 4.31).

In cases where it is necessary to locate the waler at a distance from the wall to permit wall forming, care must be taken to deal with all forces being exerted. Lateral loads being carried through the waler cannot be dissipated along the wall through the stubs (see Figure 4.15) and must be handled with specific structural details.

Corner braces can also be located directly below the waler or directly to the waler. The waler under brace method allows the use of the corner brace as a template for the waler placement. Figures 4.23 and 4.25 detail connections between strut and waler which also work for corner braces.

Because corner braces are compression members, they are usually steel column sections when the unsupported length is short (see Figure 4.32). Where the corner bracing is being used as the primary method of bracing a cut, the length of the corner braces may be quite long. In cases such as this, pipe sections (see Figure 4.33) are used for corner bracing with intermediate supports as necessary.

FIGURE 4.28 Bracing sets ready for introduction into trench shoring, Everett, WA. (Courtesy of Hurlen, Inc. Seattle, WA)

FIGURE 4.29 Strut mounted against waler with grouted connection, Colma, CA. *(Courtesy of Condon-Johnson & Associates, Inc. Oakland, CA)*

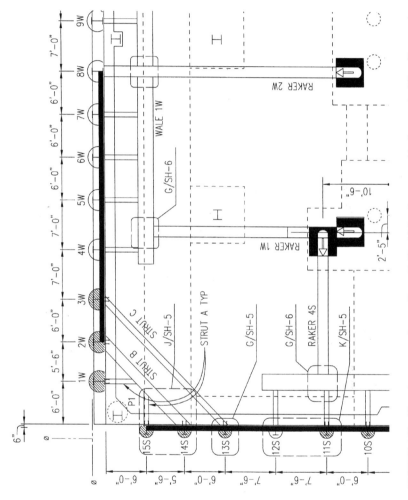

FIGURE 4.30 Plan view—corner braces (Struts B, C) with bracing mounted in flanges of soldier piles to resist lateral load. (*Courtesy of KPFF Consulting Engineers, Seattle, WA*)

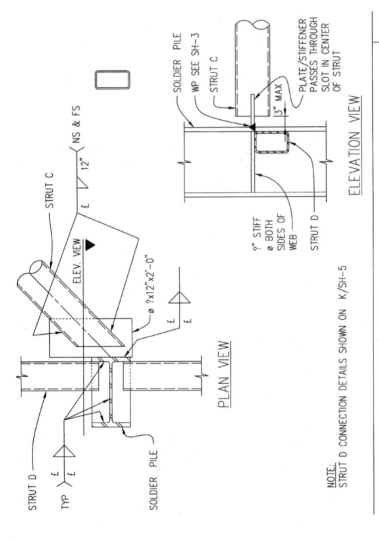

FIGURE 4.31 Detail of pipe corner brace connection. Note the square tubing in soldier pile flange. (*Courtesy of KPFF Consulting Engineers, Seattle, WA*)

FIGURE 4.32 Typical corner brace with brace mounted against waler, Issaquah, WA. (Courtesy of Condon-Johnson & Associates, Inc. Seattle, WA)

FIGURE 4.33 Internal bracing of excavation utilizing pipe corner braces, San Francisco, CA. *(Courtesy of Condon-Johnson & Associates, Inc. Oakland, CA)*

Where a shored excavation is performed to permit the construction of a building, corner braces ideally are designed to be just above a floor slab. The corner brace should be high enough above the slab to permit slab finishing below the brace, but close enough to the slab elevation to permit cutting of the corner brace as soon as the slab reaches strength. This removal transfers the brace load to the floor slab and permits further construction to progress without brace interference.

4.4 TIEBACKS

To this point, the chapter has dealt with methods of handling lateral loads by bracing within the excavation. Excavations may be supported from without the excavation by the use of tiebacks. Tieback anchors, or anchors as they are commonly called, secure the wall to a soil or rock mass which is behind that portion of the soil adjacent to the wall which is at risk of moving. See Chapter 11.4 for a discussion of the active zone.

Many methods of anchoring are available. The most commonly used methods in shored excavations are drilled and grouted anchors. However, a method, which began in the utility sector for anchoring guys and poles, involves the use of mechanical anchors. Mechanical anchor usage has spread and now they are used in lightly loaded shoring situations. Driven pipe piles have been used as tieback anchors and cases are reported of the use of driven H piles.

Anchors are almost always installed at an angle below the horizontal. This is for a number of reasons. In drilled and grouted applications, grout will run out of a horizontal hole. In driven applications, driving is much easier if it is at least at some angle of declination. In most cases, soils tend to be more competent with depth. The desire to economically use the stronger soils for anchoring is a compelling reason to install tiebacks at a downward angle.

With the exception of specific situations involving restricted right-of-way or easements, or conflicting utilities, soil anchors are usually installed at angles of between 15 and 30 degree declination to the horizontal. Rock anchors tend to be steeper in an attempt to get to rock as quickly as possible. Declination angles up to 45 degrees are common. The steepness of the angle becomes a detriment to the shoring scheme as the tieback imparts more vertical force which must be dealt with by other components of the wall.

4.4.1 Mechanical Anchors

Mechanical anchors take many forms. Two commonly used commercial anchors are helical anchors and the manta ray anchors. Helical anchors are a series of steel helical plates welded at intervals to a steel rod. The anchor is rotated into the soil with the helices literally screwing themselves into the ground. Once in place, the anchor provides pull out capacity by passive resistance (see Chapter 8.5).

Manta ray anchors are plates which are attached to a rod. The plate is advanced into the ground by impact driving. Once the plate is advanced to the depth desired, the rod is tensioned which causes the plate to rotate to a position at right angles to the rod. In this configuration, the plate provides pull out capacity through passive resistance.

An excellent reference for mechanical anchors is the *ADSC Mechanical Anchor Product Data* manual referenced in the Bibliography of this text.

4.4.2 Drilled and Grouted Anchors

Single Stage Anchors. Drilled and grouted anchors develop their pullout capacity in an entirely different fashion than mechanical anchors. These anchors mobilize the shear strength of the soil or rock by friction along their length. Figure 4.34 details various portions of an anchor. The anchor has an anchor head which attaches to the wall in order to prevent the wall from overturning. The anchor passes through an area called a "no-load zone" which is the soil which is probably subject to movement (see Chapter 11.4) and then develops its capacity in an area called the "bond zone" or "anchor zone." The anchor outlined in Figure 4.34 is what we call a single stage anchor. The top of the bond zone for all strands is the bottom of the no-load zone so that all of the strands begin developing their capacity at the same depth in the drilled hole. North American tiebacks are almost always single stage anchors.

Multistage Anchors. Drilled and grouted anchors develop their capacity by mobilizing the shear strength of the soil. Some movement is necessary in order to mobilize this shear capacity. Because the bar or strand used for anchors elongates as it is stressed, the entire load of the anchor is first brought to bear at the top of the bond zone. As the anchor elongates, the bond stresses are shed down the bond length so that the bond stresses are distributed over the length of the bond zone. This can require significant movements in the top of the bond zone in order for the stress to be uniformly distributed.

In anchors where the load is extremely high, or where the soils that the anchor is engaging are soft, the calculated anchor length can be quite long. If the entire load is placed at the top of the bond zone, the amount of movement necessary to distribute the bond stresses along the entire bond length may be so great that the soils at the top of the bond zone will fail. This phenomena can then transfer the total load further down the anchor, overloading the next segment of soil, and in a repeat of the previous occurrence, a progressive failure may occur. At a minimum, the soil/anchor bond will often be reduced to residual strength levels and optimal bond performance is not possible.

In order to overcome this problem, strand anchors can be constructed as multistage anchors (see Figures 4.35 and 4.36). With the top of the bond zone of each strand in a different place the onset of bond stresses are more evenly distributed throughout the bond zone and the soils are not overstressed in any one location.

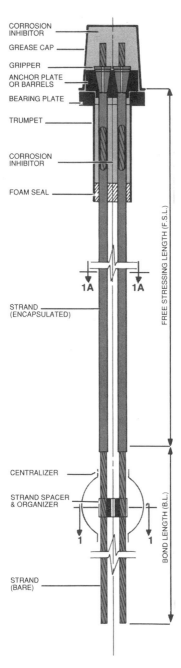

FIGURE 4.34 Simple corrosion protection-nomenclature. *(Courtesy of Con-Tech Systems Ltd. Delta, BC)*

These types of anchors are more common in Europe, but have been used in the U.S.A. One of the difficulties with this type of anchor is that each strand has a different elongation in order to achieve equal stress. Stressing must be done with multiple jacks (see Figure 4.37). This complicates the stressing operation and significantly adds to the stressing time.

4.4.2.1 Materials. A discussion of anchor components is most easily carried out if the subject of temporary anchors is addressed first with variations required to make an anchor permanent carried out later.

Temporary Anchors. Drilled and grouted anchors are constructed using either high strength steel bars (Figures 4.38 and 4.39) or post tensioning strands (Figures 4.40 and 4.41). The bars (Fs = 150ksi (1035 MPa)) are rolled with an upset thread which permits coupling and also develops bond in a manner similar to reinforcing steel bars. Bars are commercially available in diameters from 5/8 inch (16 mm) to 3½ inches (88 mm) to provide a range of capacities. Strand anchors are constructed of high strength post tensioning strands of either 0.5 or 0.6 inch (12 or 15 mm) diameter. The strand has an ultimate capacity of 270 ksi (1860 MPa) and different capacities are achieved by varying the number of strands used.

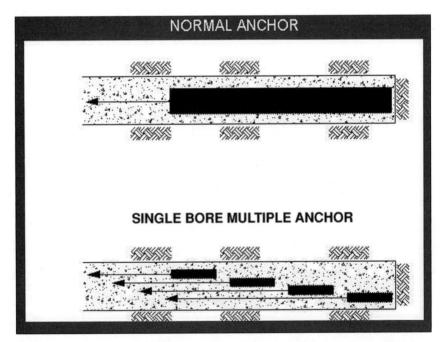

FIGURE 4.35 Schematic of multiple stage anchor. *(Courtesy of SBMA, LLC. Venetia, PA)*

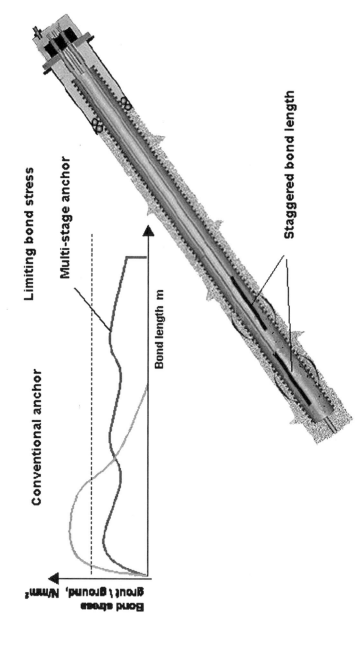

FIGURE 4.36 Multistage anchor—note change in bond stress. *(Courtesy of Dywidag Systems, Inc. Kent, WA)*

FIGURE 4.37 Stressing arrangement for multiple stage anchor. *(Courtesy of SBMA, LLC. Venetia, PA)*

Tieback anchors are positioned in the drilled hole with the use of spacers so that the grout which completes the installation will completely surround the anchor tendon. The one exception to this statement occurs when anchors are installed utilizing hollow stemmed auger techniques. No spacers are used in this application. The assembly of bar or strand together with spacers, sheathing and grout tubes is called the anchor tendon and this element is placed in the hole as one unit.

Bar anchors are attached to the wall through a plate and nut arrangement indicated in Figure 4.42. The nut is threaded so as to secure the rod to the plate. Strand anchors are attached to the wall through an anchor head and wedge arrangement. The wedges are compressed together by sliding deeper into ever decreasing strand holes machined into the anchor head.

FIGURE 4.38 Bar anchors with spacers. *(Courtesy of Dywidag Systems, Inc. Kent, WA)*

FIGURE 4.39 Bar anchors. Note the upset thread. *(Courtesy of Con-Tech Systems Ltd. Delta, BC)*

FIGURE 4.40   Single seven wire strand. *(Courtesy of Con-Tech Systems Ltd. Delta, BC)*

Strand or bar anchors are protected by a sheathing to prevent capacity being derived in the "no-load zone." This sheathing is a smooth wall polyvinyl chloride (PVC) pipe just slightly larger in ID than the OD of the bar so that the bar can slide inside the pipe. In the case of a strand anchor, the sheathing can take the form of either individual strand sheathing or one sheath which encompasses all the strands. Sheathing, which covers each strand individually, is polypropylene or high density polyethylene (HDPE) encapsulating a layer of grease which permits the strand to slide inside the sheathing (see Figure 4.43). Alternatively all bare strands may be encapsulated inside one PVC sheath.

The bond zone portion of the anchor is that portion of the anchor that comes in direct contact with the anchor grout and bonds through friction to the grout. In the case of bar anchors, a bare bar will develop bond strength through the ridges rolled onto the bar in the same fashion as a reinforcing bar bonds to concrete. Strand anchors develop their bond to the grout by friction along the length of the strand and the individual strands are spread to maximize this bond (see Figures 4.44 and 4.45).

FIGURE 4.41 Strand bundled for delivery. *(Courtesy of Con-Tech Systems Ltd. Delta, BC)*

140 EARTH RETENTION SYSTEMS

FIGURE 4.42 Anchor head details for bar anchors. *(Courtesy of Williams Form Engineering Corp. Portland, OR)*

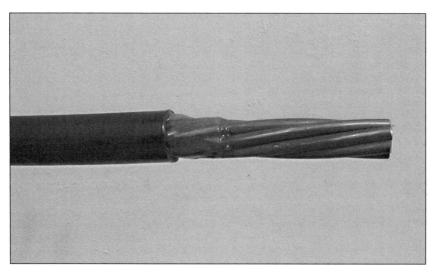

FIGURE 4.43 Greased and sheathed strand. *(Courtesy of Con-Tech Systems Ltd. Delta, BC)*

FIGURE 4.44 Bare strand with strand organizers. *(Courtesy of ADSC-The International Association of Foundation, Drilling. Dallas, TX)*

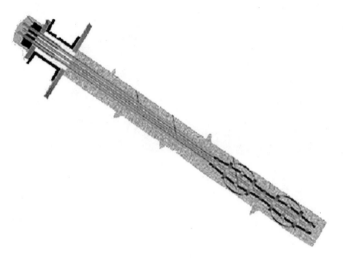

FIGURE 4.45   Strands spread with spacers—schematic. *(Courtesy of Dywidag Systems, Inc. Kent, WA)*

Permanent Anchors. In order to provide the corrosion protection required for permanent anchors, it is necessary to encapsulate the entire anchor. A complete discussion of corrosion protection of anchors is contained in *PTI Manual for Soil and Rock Anchors* (see Bibliography for reference).

Figure 4.46 details a permanent bar anchor. Note that the entire anchor is encapsulated in grout within a ribbed sheath. The ribbing of HDPE or PVC provides roughness so that the anchor can develop bond capacity with the grout on the outside of the sheath (see Figure 4.47). Note that a sheath of smooth plastic is used to prevent bond development within the no-load zone. Figure 4.48 is a photo of a bar anchor encapsulated in grout and sheathing.

A pipe of steel or plastic is attached to the back of the base plate. This pipe, called a trumpet, protects the anchor from corrosive elements as it transitions from the ribbed sheath to the anchor head. The trumpet is either filled with grout, foam, or grease, after stressing to provide complete protection. A cap filled with corrosion inhibitor is placed over the lock off nut to protect the anchor head. In the case of permanent strand anchors, the anchor zone (bond length) consists of bare strand covered with a corrugated PVC or HDPE sheathing. The sheathing is filled with grout. The no-load zone consists of individual strands greased and sheathed in a smooth PVC casing. The strand bundle is then placed inside a smooth wall PVC or Polyethylene casing which covers the no-load zone portion of the tendon. This casing is filled with grout to provide corrosion protection.

Similar to a bar anchor, a trumpet protects the strand anchor as it transitions from its PVC casing protection to the anchor head. Figure 4.49 details a corrosion protected strand anchor. Figures 4.50 and 4.51 display epoxy coated anchors that are an alternative used for corrosive environments.

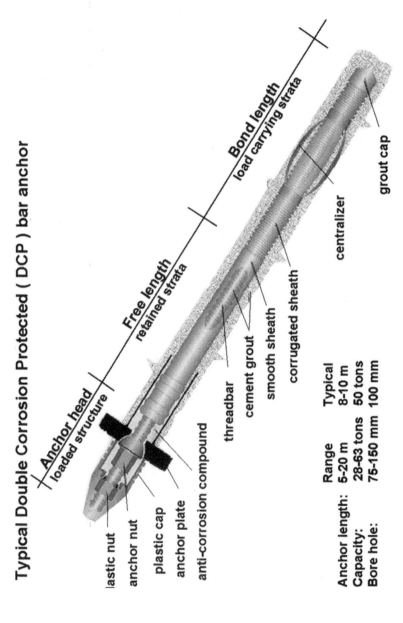

FIGURE 4.46 Typical permanent bar anchor. *(Courtesy of Dywidag Systems, Inc. Kent, WA)*

FIGURE 4.47 Stockpiled corrugated casing. *(Courtesy of Con-Tech Systems Ltd. Delta, BC)*

FIGURE 4.48 Encapsulated bar for permanent anchoring. (*Courtesy of Dywidag Systems, Inc. Kent, WA*)

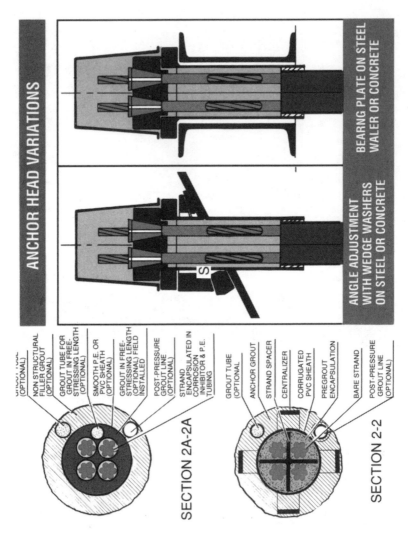

FIGURE 4.49 Typical permanent strand anchor. (*Courtesy of Con-Tech Systems Ltd. Delta, BC*)

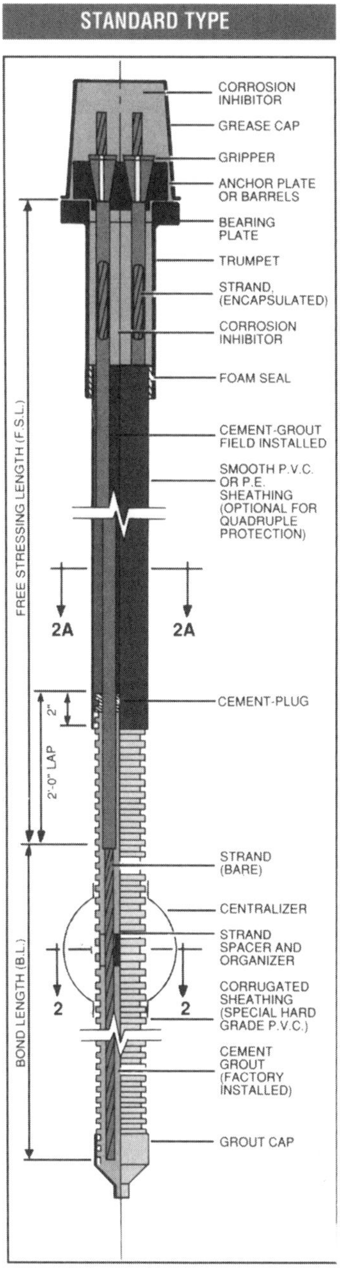

FIGURE 4.49 *(continued)* Typical permanent strand anchor. (Courtesy of Con-Tech Systems Ltd. Delta, BC)

FIGURE 4.50 Epoxy coated bar used for soil nails or anchors. *(Courtesy of Condon-Johnson & Associates, Inc. Seattle, WA)*

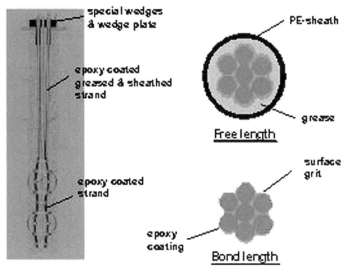

FIGURE 4.51 Typical epoxy coated strand anchor details. *(Courtesy of Dywidag Systems, Inc. Kent, WA)*

4.4.2.2 Installation. Drilled and grouted anchors can be installed in a number of ways. The equipment used is outlined in Chapter 13. The holes may be drilled utilizing auger rigs or continuous flight augers. Holes drilled utilizing this method range in size from 8 to 30 inch (200-760 mm) diameter. The anchor tendon is usually placed prior to grouting, although some instances have been noted of higher bond capacity being developed by installing the tendon after grouting (called "wet setting"). Grout is poured into dry holes or tremied into wet holes.

Anchors can be installed by hollow stemmed continuous flight augers in a method called auger casting. An anchor tendon is placed inside the auger and the auger drilled into the ground. Once the auger reaches design depth, grout is forced down the hollow stem of the auger and the auger is withdrawn leaving the grout and tendon in place. These augers range in size from 8 to 18 inch (200-460 mm) diameter.

Anchors can also be installed by rotary techniques utilizing air or water as a flushing medium. This method utilizes drag bits, rotary bits with top hole percussion hammers, or down hole hammers to drill the hole. Once the hole is completed, the drill string is withdrawn and a tendon set and grouted in place. Hole sizes are in the range of 4-10 inches (100-250 mm) in diameter.

Duplex drilling techniques are commonly used in tieback drilling. In this method a hole is advanced by rotary techniques. The hole is protected by a casing which is advanced simultaneously with the drill bit. Hole sizes are in the range of 5-8 inches (125-200 mm). Once hole depth is reached, the drill string and bit are withdrawn but the casing remains. The tendon is then placed in the hole and grouting begins. As grouting continues, the casing is withdrawn.

4.4.2.3 Grouting. Grouting is usually performed with neat cement grouts. Bagged or bulk cement is mixed with water on site at a rate of 5-6 gallons (19-23 L) per sack of cement. This grout is then pumped down the drilled hole through 1 inch (25 mm) diameter lines.

Commercially purchased and delivered grouts can be used provided that pressure grouting is not required. The use of sand grouts is typically seen in auger and continuous flight auger tiebacks. These grouts usually are mixed at a ratio of 9 sacks (385 kg) of cement per cubic yard (0.765 m) with sand aggregate. No coarse aggregate is used as it does not pump well in the 2 inch (50 mm) diameter lines commonly used.

Grouting under pressure has been found to significantly increase the bond strength between the grout and soil. Two grouting methods are commonly used: pressure grouting and secondary grouting.

Pressure grouting is performed in a cased hole and consists of pumping grout under pressures of up to 150 psi (1 MPa). The grout can be pumped under pressure because it is pumped through a cap that is attached to the top of the casing. Once pressure is attained, the cap is removed and one casing length (usually two meters) is removed. The process is repeated until the bond zone of the tieback is pressure grouted.

Secondary grouting is performed after the hole has been initially grouted (primary grouting) and the grout has taken its initial set. The anchor tendon is made up with a secondary grout line which leads to a series of grout valves. This grout line has a return line to the surface. The primary grout is introduced to the hole by gravity methods. Once the grout has set (usually 24 hours), water is pumped through the secondary grout lines. With the return line sealed, pressure is applied to open the grout valves and fracture the initial grout (this method is also called fracture grouting). Pressures as high as 800-1000 psi (5.5-6.9 MPa) may be required to open the grout valves. Once the grout valves are open, grout is pumped through the valves to form high pressure grout balls which significantly increase the anchor capacity.

If pressure cannot be held at satisfactory levels during secondary grouting, it may be necessary to terminate grouting and perform the operation again after this round of grouting has set. In order to do this, the return line is opened and water is pumped down the secondary grout line and out the return line to flush any grout out of the grout pipes. Secondary grouting is then repeated until satisfactory pressure can be maintained.

4.4.2.4 Stressing. All anchors are stressed as part of a quality control program (see Chapter 14). Anchors are tested either for verification, ensuring that the design assumptions and techniques are correct; performance, ensuring that design methods continue to be appropriate for conditions found in the field; or proof, ensuring that specified techniques are being adhered to and capacities are being achieved.

4.4.2.5 Attachment Techniques. Anchors are attached to shoring walls in a number of methods. In the case of soldier piles or secant piles, anchors can be attached directly to the pile. Figures 4.52 through 4.58 detail several techniques used. Where the flange is cut to permit passage of the tieback close to the pile web, a cover plate is placed opposite the tieback to replace the lost pile flange and restore section. This technique can produce torsion in the soldier pile as an eccentricity exists between the anchor and the pile web. Tight tolerance control must be adhered to. Alternatively the torsion can be dissipated by placing a waler between the piles (see Figures 4.59 and 4.60) or by strapping the pile under torsion to the adjacent pile (see Figures 4.61 and 4.62). The direct connection cannot be used for driven soldier piles unless the connection is fabricated after driving.

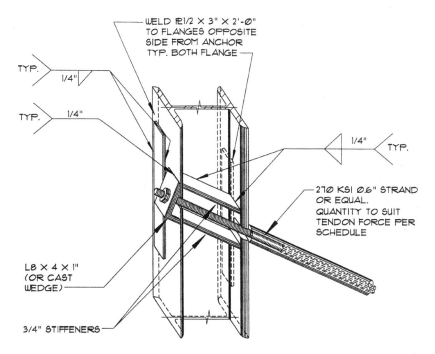

FIGURE 4.52 Typical detail—direct connection of tieback to pile. *(Courtesy of CT Engineering Inc. Seattle, WA)*

FIGURE 4.53 Direct tieback to pile connection—fabricated bearing seat, Seattle, WA. (*Courtesy of ADSC-The International Association of Foundation. Drilling Dallas, TX*)

FIGURE 4.54 Direct connection of tieback to pile—pipe seat, Seattle, WA. *(Courtesy of Condon-Johnson & Associates, Inc. Seattle, WA)*

FIGURE 4.55 Direct connection of tieback to pile—angle seat, Syracuse, NY. *(Courtesy of Isherwood Associates, Oakville, Ont.)*

FIGURE 4.56 Note web stiffeners for direct soldier pile to tieback connection, Bellevue, WA. *(Courtesy of Condon-Johnson & Associates, Inc. Seattle, WA)*

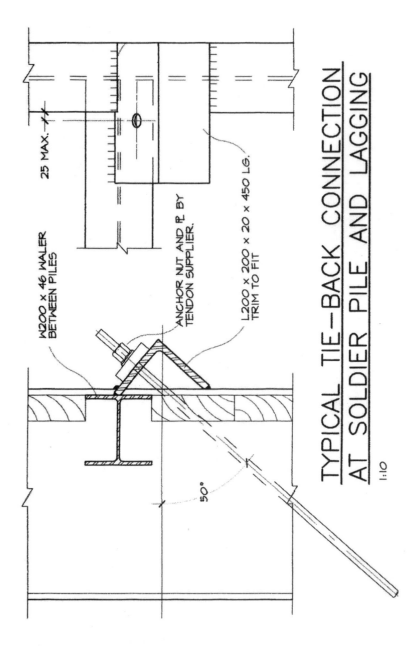

FIGURE 4.57 Alternative detail for connection of soldier pile to anchor. (*Courtesy of Isherwood Associates. Oakville, Ont.*)

FIGURE 4.58 Direct connection of pile to tieback, Toronto, Ont. (see 4.58). *(Courtesy of Deep Foundations Contractors, Thornhill, Ont.)*

FIGURE 4.59 Tieback connection with flush mounted waler, Toronto, Ont. (*Courtesy of Isherwood Associates. Oakville, Ont.*)

FIGURE 4.60 Connection of pile to anchor using flush mounted waler, Toronto, Ont. (*Courtesy of Deep Foundations Contractors. Thornhill, Ont.*)

FIGURE 4.61 Strapping used to relieve torsion in soldier pile from anchor load-corner condition, Toronto, Ont. (*Courtesy of Isherwood Associates. Oakville, Ont.*)

FIGURE 4.62 Strapping used to relieve torsion in soldier pile from anchor load, Seattle, WA. Setback lagging causes lack of lateral support of front flange of pile encouraging torsional problems. *(Courtesy of Condon-Johnson & Associates, Inc. Seattle, WA)*

Soldier piles fabricated from two wide flanged sections or two channels have been constructed which permit the placement of tiebacks through the center of the section. This arrangement detailed in Figures 4.63 through 4.65 completely eliminates problems with torsion. These piles cannot be driven and must be placed in drilled holes. Fabricated double piles are much more expensive than single pile sections.

Walers can be constructed to span from soldier pile to soldier pile. These walers can receive either one tieback in the center of the span between piles or two tiebacks, one beside each soldier pile. These walers can be constructed from H-Pile sections, with a cutout in the center to permit passage of the tieback (Figures 4.66 and 4.67), back to back channels (Figures 4.68 and 4.69), back to back wide flange beams (Figure 4.70) or square tubing (Figure 4.71). Walers of this type of construction are always mounted normal to the tieback tendon and must be mounted on some form of wedge to bring about this alignment.

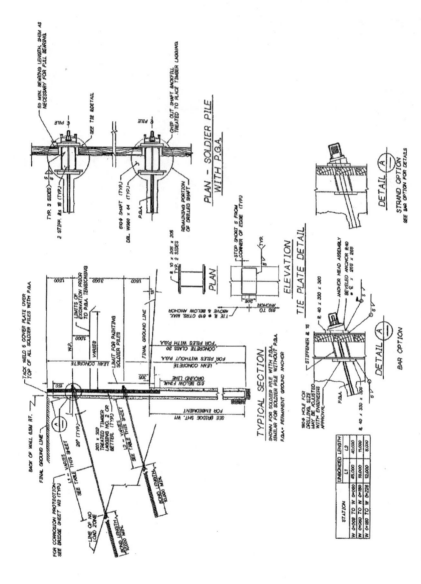

FIGURE 4.63 Typical soldier pile utilizing doubled wide flange beams.

FIGURE 4.64 Double soldier pile, Hobart, WA. *(Courtesy of Condon-Johnson & Associates, Inc. Seattle, WA)*

FIGURE 4.65 Tieback being drilled through a paired channel soldier pile, Mercer Island, WA. (*Courtesy of ADSC-The International Association of Foundation Drilling, Dallas, TX*)

LATERAL SUPPORT 165

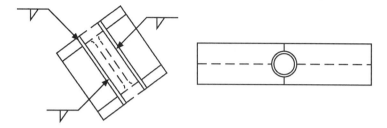

FIGURE 4.66 Typical section—H beam waler.

FIGURE 4.67 Beam waler between soldier piles, Bradford, PA.

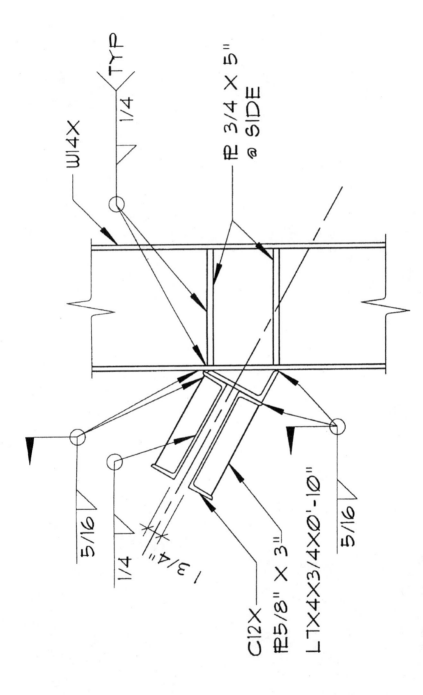

FIGURE 4.68 Typical section—double channel waler. (*Courtesy of CT Engineering, Inc. Seattle, WA*)

FIGURE 4.69 Double channel waler, Seattle, WA. *(Courtesy of Condon-Johnson & Associates, Inc. Seattle, WA)*

FIGURE 4.70 Double beam waler, Toronto, Ont. *(Courtesy of Deep Foundations Contractors. Thornhill, Ont.)*

FIGURE 4.71 Square tube waler, Seattle, WA. *(Courtesy of Condon-Johnson & Associates, Inc. Seattle, WA)*

Walers can also be constructed as cast-in-place concrete beams attached to the soldier piles (see Figure 4.72).

Tiebacks are most often attached to sheet pile walls utilizing channel wales. Slurry wall tiebacks are usually attached directly through the wall with reinforced blockouts either poured in the wall section or precast as separate blocks.

4.5 DEADMEN ANCHORS

When anchoring is possible at a very shallow level which will adequately provide overturning resistance to a shoring wall, deadman anchors should be considered. These anchors are similar to tieback anchors except that in order to develop their capacity they are attached to some form of buried anchorage which will resist movement through mobilization of passive pressures.

Deadman anchors are installed as horizontal anchors. The tendons can be bar or strand and they can be treated for corrosion exposure in a manner similar to drilled and grouted anchors (see *PTI Manual for Soil and Rock Anchors*—reference in Bibliography).

The anchorage, called a deadman, can take various forms. It may be a wall of short driven sheet piles. It may also be a buried precast concrete anchorage, or it could be a continuous cast-in-place concrete beam.

If the anchored wall is comprised of soldier piles, the attachment of a deadman anchor can be by direct attachment as detailed in Section 4.4.2.5. Soldier pile walls and sheet pile walls can also utilize walers similar to those detailed in Section 4.4.2.5. Most often used is the double channel connection. If the channel is applied to the outside of the shoring wall with the anchor tendon passing through the wall, the connection of waler to pile is in compression and the connection is very simple. If the waler is attached to the backface of the shoring wall (see Figure 4.73) the attachment will be in tension and a weldment or bolted arrangement must be designed to deal with these loads. The waler-behind-wall connection yields a much cleaner face for the shoring wall.

Figure 4.74 details a variation of deadman anchoring. This pier utilizes tendons which connect to the wall on the other side of the pier. In effect, each wall acts as a deadman for the other. Figure 4.75 details a concrete deadman which will be buried in the subsequent fill to provide anchorage.

Not all deadman anchorages are installed in a mass excavation as indicated in Figures 4.73 through 4.75. A deadman can be installed in a trench dug parallel to the shoring wall at the proper distance behind the wall. The tendons are then brought through to the deadman for connection either by cutting small cross trenches or by horizontal drilling.

FIGURE 4.72 Cast-in-place concrete waler, Cloverdale, CA. *(Courtesy of Condon-Johnson & Associates, Inc. Oakland, CA)*

FIGURE 4.73 Double channel waler mounted behind sheet piling for connection to deadman. Note bolting being used for tension connection. (*Courtesy of Dywidag Systems, Inc. Kent, WA*)

FIGURE 4.74 Cross ties make each wall a deadman for the other. *(Courtesy of Dywidag Systems, Inc. Kent, WA)*

FIGURE 4.75 Deadman connection to cast-in-place concrete deadman. *(Courtesy of Dywidag Systems, Inc. Kent, WA)*

4.6 CANTILEVER SHORING

Soldier pile, secant pile, cylinder pile, sheet pile, and slurry walls can all be designed within limits to stand without any component of lateral restraint other than their own embedment (Figure 4.6). Chapter 11.1 will deal with design methods to effect this result. Figure 4.76 details a cantilever pile and Figure 4.77 details a typical soldier pile and lagging cantilever wall.

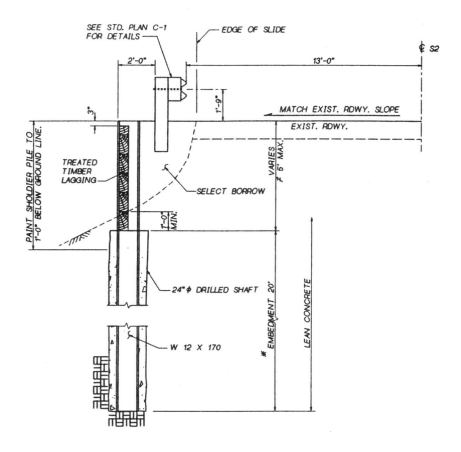

FIGURE 4.76 Typical cantilever soldier pile and lagging. *(Courtesy Washington State Department of Transportation)*

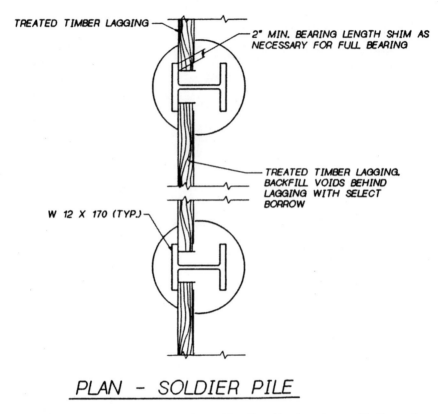

PLAN - SOLDIER PILE

FIGURE 4.76 *(continued)* Typical cantilever soldier pile and lagging. *(Courtesy Washington State Department of Transportation)*

4.7 SOIL NAILS

Until now, this chapter has dealt with lateral earth pressure by restraining the face of the shoring wall. The soil nailing technique reinforces a soil mass and strengthens it so that the soil will act as a block. This is done by the installation of regular inclusions called soil nails.

Although totally different in their operation, soil nails look, for all intents and purposes, like soil anchors. In most applications, a soil nail consists of a reinforcing steel bar ranging from # 7 to #10 (#22-#32) in size and grading from either regular rebar grades (60 or 75 ksi (415-520 MPa)) to high strength (150 ksi (1035 MPa)) centered in a hole of 6 to 8 inch (150-200 mm) in diameter which is filled with high strength grout (see Figure 4.78). Nails may also be comprised of hollow steel rods. These rods act as sacrificial drill steel, and are drilled into

FIGURE 4.77 Cantilever soldier pile retaining wall, Shoreline, WA. *(Courtesy of Condon-Johnson & Associates, Inc. Seattle, WA)*

the ground and grouted using the center hole as a grout channel (see Figure 4.79). Some work has been done with split sets (see Figure 4.80), driven nails or nails fired under air pressure, but by far the majority of nails in North America are installed by drilling and grouting.

The majority of soil nails are installed utilizing gravity grouting techniques Some recent work, in softer soils, has incorporated secondary grouting techniques also described in Section 4.4.2.3.

Once the nails in a particular lift are installed, a facia of shotcrete is applied to cover the exposed soil face. This fascia is attached to the nails by plates which are captured on the ends of the nails with nuts (see Figures 4.81 and 4.82). In cases where the soil nail is deemed to be permanent, it is attached to the completed structure by way of a studded plate (see Figures 4.83 and 4.84).

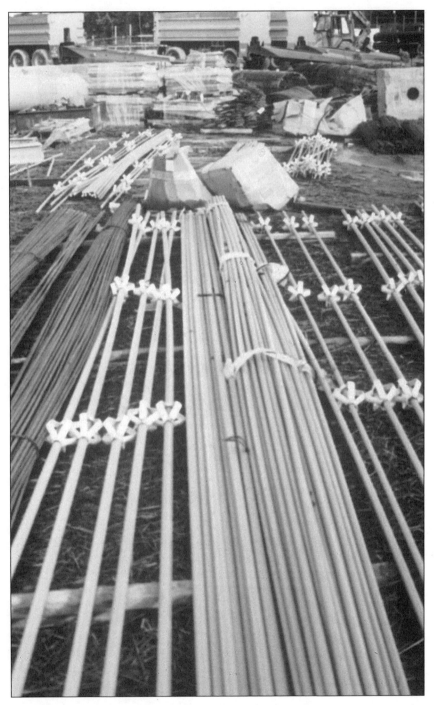

FIGURE 4.78 Epoxy coated permanent nails with spacers. *(Courtesy of Golder Associates Inc. Redmond, WA)*

FIGURE 4.79 IBO-BAR—used for self drilled soil nails or anchors. (Courtesy of Con-Tech Systems Ltd. Delta, BC)

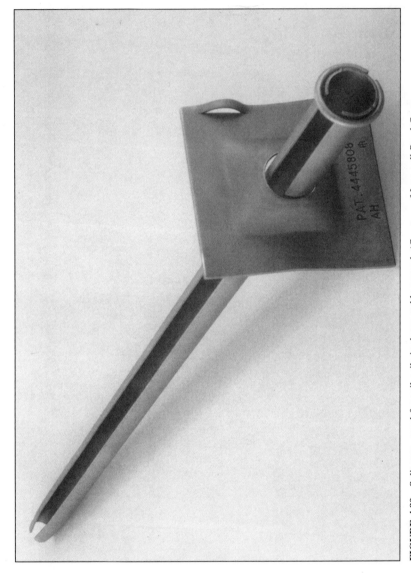

FIGURE 4.80 Split sets—used for soil nails in dense stable ground. (*Courtesy of Ingersoll-Rand Company. Roanoke, VA*)

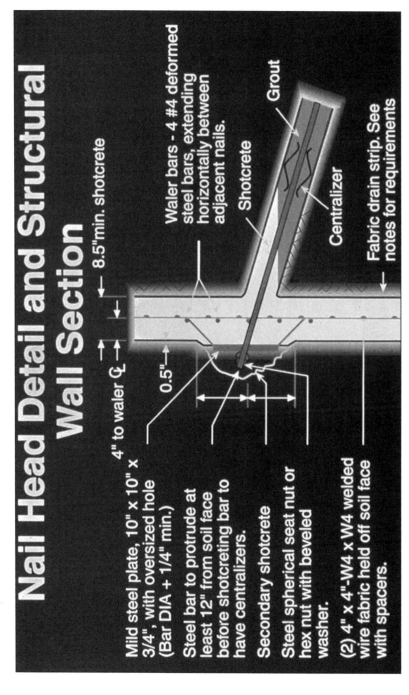

FIGURE 4.81 Typical soil nail and plate detail. (*Courtesy of Golder Associates Inc. Redmond, WA*)

FIGURE 4.82 Nail and Plate prior to shotcrete application, Redmond, WA. *(Courtesy of Condon-Johnson & Associates, Inc. Seattle, WA)*

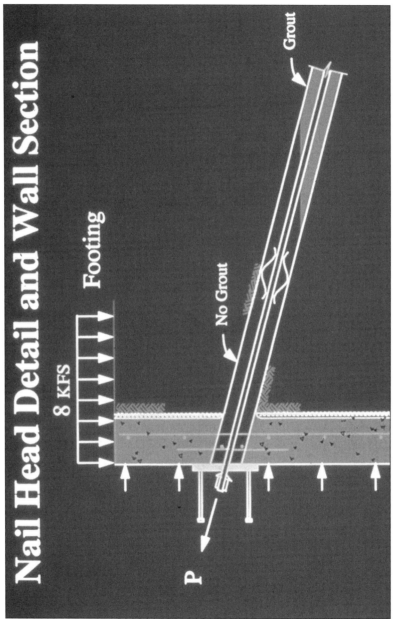

FIGURE 4.83 Typical soil nail and studded plate detail. *(Courtesy of Golder Associates Inc. Redmond, WA)*

FIGURE 4.84 Soil nail and studded plate, Portland, OR. *(Courtesy of Condon-Johnson & Associates, Inc. Seattle, WA)*

LATERAL SUPPORT 185

Permanent soil nails are usually protected from corrosion by epoxy coating the nail. See *FHWA Manual For Design and Construction Monitoring of Soil Nailed Walls* (reference in Bibliography). In some cases, part or all of the nail may be encapsulated in grout and ribbed PVC sheathing similar to permanent soil anchors (Section 4.4.2.1).

4.7.1 Strut Nails

The shotcrete fascia can sometimes place a very heavy load on the ends of the nails which must act in cantilever to support the weight. This occurs most often in situations where shotcrete is being installed in a permanent application as described in Chapter 6.2. Thicknesses of shotcrete up to 24 inches (610 mm) have been used in heavy retaining wall cases. When the shotcrete is too heavy for the nails to carry in cantilever (usually greater than 10 inches (250 mm) in thickness in the first one or two lifts), the engineer can overcome this problem by adding strut nails. These short nails (usually 10 feet (3.0 m) in length) are very steeply inclined (70 degrees to the horizontal). Acting as short micropiles, they carry the load of the shotcrete in compression until sufficient lifts of shotcrete are in place to mobilize wall friction.

CHAPTER 5
FACING

Most of the systems discussed in Chapter 3 have fascia elements that are integral with the primary vertical elements. Sheet piling presents full face coverage. Secant piles and tangent piles cover the entire excavated face with the concrete placed in the excavated shaft. Slurry walls present a complete face of tremied concrete and Trench Boxes incorporate the facing panels as an integral part of the box.

Three systems have separate fascia systems. These are soldier pile and lagging, soil nailing, and micropile walls. Underpinning in Chapter 3 is often actually an adaptation of soldier pile and lagging and so will not be dealt with separately. In soldier pile, and lagging and micropile walls, the fascia is called lagging while the fascia of soil nailed systems is a thin shell placed by shotcrete methods.

5.1 LAGGING

The word lagging, as it is used it in the earth retention industry, has nothing to do with the facing on a hoisting drum, nor the habit of falling behind. Lagging, in this context, describes the material used to span the gap between soldier piles. While it is usually wood, and placed by hand, it does not necessarily have to be so. It can be of concrete or steel. The span between soldier piles is normally in the range of 6-10 feet (1.8-3.0 m). Soldier pile and lagging systems are designed as free draining systems, so that any water which encounters a lagged wall is expected to seep through the wall. Timber lagging is therefore ideal as it permits flow between the planks and is manageable as a manual load.

Lagging can only be installed in materials that demonstrate some stand up capability. In other words, a face of soil must be exposed for some period of time in order to install the lagging boards. During this time, the face must remain stable. Lagging is usually installed in lifts of approximately four to five feet (1.2-1.5 m). The planks in each lift are installed from the bottom up. Once a lift is complete, the lift is secured to the piling by wedging or nailing to prevent it from slipping during further excavation. The next lift is then excavated and the process repeated.

Lagging can be utilized in either temporary or permanent applications. Temporary applications usually are for periods of less than one year and occur when soldier pile and lagging is used as temporary excavation support for construction of buildings, utilities or civil engineering installations. Permanent applications occur when lagging is the final exposed fascia for retaining walls constructed utilizing soldier piles.

Lagging can be either tucked between the flanges (see Figures 5.1 and 5.2) or mounted on the face of the soldier piles (see Figures 5.3 and 5.4). On occasions where it is not possible to place the soldier pile in a location that will permit placement of the lagging behind the front flange, the lagging can be blocked back behind the front flange of the pile with either timber blocking or welded angle clips (see Figure 5.5). Lagging can even be placed behind the back flange of the soldier pile.

5.1.1 Material

5.1.1.1 Timber. Timber, the most commonly used lagging material, can be a variety of species. On the West Coast, lagging is usually Douglas fir or Hem-fir. On the East Coast, and in the South, mixed hardwood is used. Timber lagging is usually 3, 4 or 6 inches (75, 100 or 150 mm) thick and is generally full dimension thickness. In other words, unlike dressed lumber where nobody knows the real dimension of a 2 x 4, but everyone knows that it isn't 2 inches by 4 inches, 4 inch (100 mm) timber lagging is actually 4 inches (100 mm) thick. Lagging planks are usually supplied in widths of 8 to 12 inches (200-300 mm). The thicker the plank, the narrower the width in order to keep the weight of the plank manageable for lifting.

Some lagging is sold which is cut slightly less than the advertised dimension. These planks are sometimes referred to as "scants" and the amount of undersizing appears to be equal to the saw thickness. While there is nothing wrong with lumber which is slightly less than advertised dimension, the effect of the undersizing should be considered when specifying the lagging (Chapter 11.8).

Timber lagging can be impregnated with treatments such as CCA (chrominated copper arsenate) for Hem-fir, ACZA (ammoniacal copper zinc arsenate, called Chemonite) for Douglas fir, or pentachlorophenol for both. Mixed hardwoods can be treated with creosote. When its use is permanent, it almost always is treated. Some municipalities require that temporary lagging be treated, but these are in the minority.

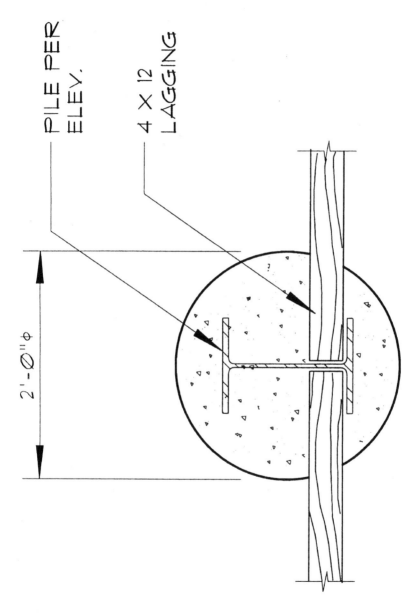

FIGURE 5.1 Lagging placed behind front flange of soldier pile–schematic. (*Courtesy of CT Engineering, Inc. Seattle, WA*)

FIGURE 5.2 Lagging placed behind front flange of soldier pile, Seattle, WA. *(Courtesy of Condon-Johnson & Associates, Inc. Seattle, WA)*

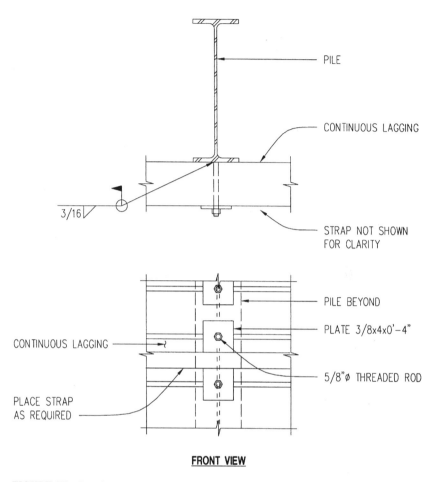

FIGURE 5.3 Lagging mounted on face of soldier pile with threaded rod and clip attachment—schematic. *(Courtesy of KPFF Consulting Engineers. Seattle, WA)*

With the exception of steel "road" plates which are discussed later, timber lagging provides the greatest flexibility when dealing with standup time. Standup time is a measure of the amount of time that an exposed soil face will stand prior to the onset of raveling. In cohesive soils it is almost never a problem, but it must be considered in cohesionless soils. Sands and gravels which have a significant silt fraction, or have some form of cementing in their structure usually do not experience standup time problems. Those sands and gravels which do not have cementing or are not sufficiently silty may still have good stand up time characteristics because of apparent cohesion (Chapter 8.2.1). The evaluation of the potential for a soil face to stand, especially when apparent cohesion is being

FIGURE 5.4 Face mounted lagging, Houston, TX. *(Courtesy of Schnabel Foundation Co. Inc. Houston, TX)*

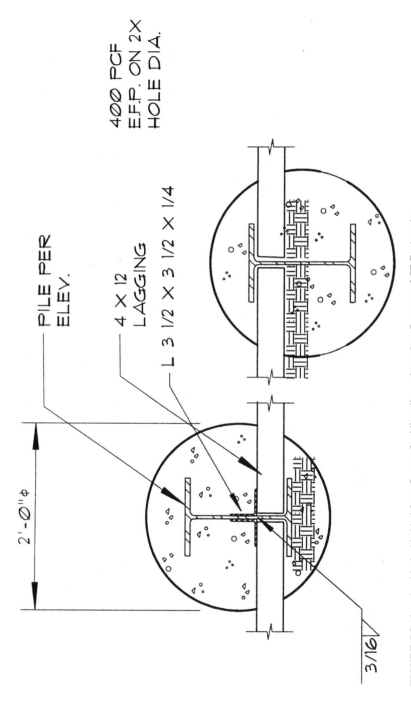

FIGURE 5.5 Lagging blocked behind front flange of soldier pile—schematic. (*Courtesy of CT Engineering, Inc. Seattle, WA*)

relied upon, is very much an exercise in observation of actual test pits or the application of previous experience. Planks are cut to fit utilizing chain saws. The standup time required to cut and clean the face and place the planks can, if necessary, be reduced to something in the neighborhood of one-half hour.

5.1.1.2 Concrete. Shotcrete and precast lagging are two concrete applications used. Precast lagging is only used in permanent applications. Precast lagging (Figure 5.6) can be visually very appealing, and patterns can be cast into the lagging to increase its aesthetics. While precast lagging certainly solves the problem of long term deterioration which might affect treated lagging, it requires very tight tolerances when installing soldier piles. Precast lagging cannot be cut to fit as readily as timber for placement between the soldier pile flanges and jobsite timing usually requires that lagging be cast prior to soldier pile installation.

On the other hand, shotcrete is used in both temporary and permanent applications. Temporary shotcrete, usually in thicknesses of 4-5 inches (100-125 mm), is reinforced with mesh. Permanent lagging is somewhat thicker and reinforced with reinforcing steel. Drainage, which occurs in timber or precast lagging through the plank joints, is provided by placing drain fabric on the excavated soil face prior to the placement of the shotcrete. Frequent drain holes through the shotcrete allow water to be relieved from the drain fabric.

Shotcrete lagging (see Figure. 5.7) can be placed behind the front flange of the pile or attached to the face of the pile by the use of studs welded to the soldier pile (see Figure 5.8). Shotcrete placed as a facing on micropile walls will always be attached by use of studs. Patterning of shotcrete is not as simple as it is with precast lagging and certainly the contouring of shotcrete can add significantly to its cost. See Chapter 5.2 for a discussion of shotcrete as a fascia.

Standup time becomes more important when dealing with concrete lagging. It is virtually impossible to install precast lagging in multiple lifts. The soil must stand while excavated to the full depth of the cut to permit placement of the lagging from the bottom up. Alternatively, it may be necessary to place timber lagging behind the back flange of the soldier piles during excavation to provide stability. Once the base of the excavation is attained, the precast lagging can be spaced between the soldier pile flanges.

Standup time is similarly important in the case of shotcrete. By the time the soil face is cut and trimmed, drainage fabric placed, reinforcing mesh hung and shotcrete applied, a minimum of four (4) hours has elapsed.

5.1.1.3 Steel. While not nearly as common as timber, steel has been used as a lagging substance. Metal decking has been used in places where it was felt that the long-term deterioration of lagging might be detrimental to adjacent building footings (see Figure 5.9). The decking is placed between soldier piles and grout is pumped behind it to fill voids.

FIGURE 5.6 Precast lagging, White Pass, WA. (Courtesy of Condon-Johnson & Associates, Inc. Seattle, WA)

FIGURE 5.7 Shotcrete lagging, Seattle, WA. *(Courtesy of Condon-Johnson & Associates, Inc. Seattle, WA)*

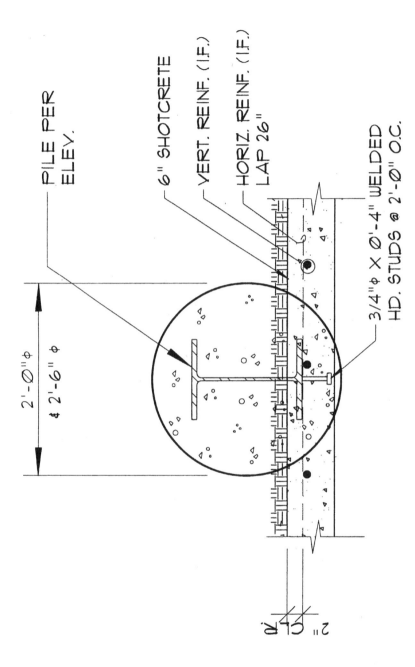

FIGURE 5.8 Shotcrete lagging—schematic. (*Courtesy of CT Engineering, Inc. Seattle, WA*)

FIGURE 5.9 Steel deck lagging, Seattle, WA. *(Courtesy of Condon-Johnson & Associates, Inc. Seattle, WA)*

In cases where the soils consist of loose to medium dense sands without cobbles or boulders, contractors have driven steel "road" plates between the soldier piles (see Figure 5.10). This system has the advantage of providing shoring protection without exposing the soil face to standup tme issues. The sheets are driven utilizing a vibro hammer and extracted for reuse. Provided obstructions are not encountered, it is a very economic way to install lagging to depths of up to 20 feet (6.1 m). The soldier pile must be placed very accurately for both plumb and location in order for the system to work satisfactorily.

5.1.1.4 Plastic. A proprietary lagging is produced called Dura-Lagg. It is made up of hollow planks made from reclaimed plastic which are very easy to move about in tight quarters and difficult access projects. Reinforcing rods can be added just prior to placing of the plank and the plank cavity is filled with cement grout after installation of the plank (see Figure 5.11).

5.2 SHOTCRETE FASCIA

Shotcrete fascias on soil nail systems carry out two responsibilities. Firstly, they provide weather protection so that slaking and drying do not rob the face of its ability to stand. Secondly, they handle any loads which are exerted at the face. Theoretically, the fascia of a soil nailed system experiences no lateral load. Experience and many field measurements indicate that some load is evident at the face. The load on the shotcrete fascia approaches 30 percent of that which you might predict by using Rankine or Apparent Earth Pressure analyses. That load is sufficient that it must be taken seriously when designing a shotcrete fascia.

Shotcrete fascias for temporary soil nailing are generally 4 inches thick and reinforced with a light mesh similar to that used for slab on grade construction. A 4 x 4 (100 x 100 mm), W2.9 x W2.9 mesh is usually adequate to permit the shotcrete to span between nails. In addition, the shotcrete must deal with the concentration of load around the nails. This is done with the use of waler bars, tic-tac-toe bars and plates.

Waler bars are horizontal reinforcing steel bars that form a sort of light horizontal beam through the shotcrete. Usually waler bars consist of 2 x 4's (#13) running horizontally across each row of nails (see Figure 5.12).

Tic-Tac-Toe bars are reinforcing steel bars which spread the shear stresses in the shotcrete. They usually consist of 2 x 4's (#13) x 3 feet (915 mm) each way under the nail plate (see Figures 5.13 and 5.14).

Drainage strips made of dimpled plastic and filter fabric (see Figure. 5.15) are installed vertically at 6 feet cc (1.83 m) and cross linked between shotcrete lifts to provide drainage behind the shotcrete fascia. Drain strips are normally 12-16 inches (300-450 mm) wide. Drain strips can be seen in Figures 5.12 and 5.14.

FIGURE 5.10 Street plates used as lagging, San Diego, CA. *(Golder Associates Inc. Redmond, WA)*

FIGURE 5.11 Dura-Lagg patented lagging system utilizing a plastic form which is reinforced and filled with grout. (*Courtesy of Don Morin, Inc. Sumner, WA*)

FIGURE 5.11 *(continued)* Dura-Lagg patented lagging system utilizing a plastic form which is reinforced and filled with grout. *(Courtesy of Don Morin, Inc. Sumner, WA)*

FIGURE 5.12 Waler bars. Note the horizontal rebar on either side of soil nail blockout, Portland OR. *(Courtesy of Condon-Johnson & Associates, Inc. Seattle, WA)*

204 EARTH RETENTION SYSTEMS

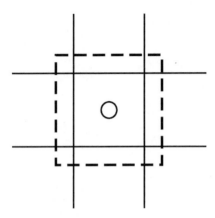

FIGURE 5.13   Typical tic-tac-toe reinforcing for soil nails.

FIGURE 5.14   Tic-tac-toe reinforcing prior to shotcrete application, Los Angeles, CA. *(Courtesy of Dywidag Systems, Inc. Kent, WA)*

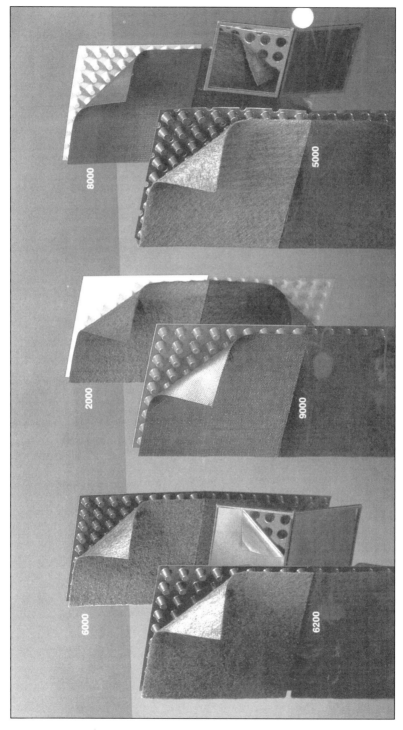

FIGURE 5.15 Drain fabric used with soil nailing to drain walls. *(Courtesy of T.C. Miradri Corp., Norcross, GA)*

Each nail has a plate which concentrates the wall loading onto the nail. Typical plates are 8 inches x 8 inches x ½ inch (200 x 200 x 12 mm) and are captured by a nut sitting in a coned washer.

Permanent exposed shotcrete fascias must conform to the design parameters of permanent basement walls. As a result, permanent walls are 8 inches (200 mm) or thicker and have at least one mat of reinforcing steel included (see Figure 5.16). In view of this steel inclusion, the requirement for waler bars and tic-tac-toe bars is eliminated. In order to handle the build up of stresses at the nail head which eventually occurs in permanent wall/permanent nail situations, the nail plates are studded and embedded in the permanent wall.

5.2.1 Vertical Elements

Soil nail systems which feature shotcrete fascias (about 95 percent of the applications) rely very heavily on the excavation of a stable face against which to shoot the shotcrete. In cases where the soil standup time is marginal or where sloughing is exacerbated by the exposure time of the cut face, the installation can be improved with the use of vertical elements. These elements, used to increase facial stability, are drilled vertical holes of 6 inch diameter (150 mm). The holes are installed and grouted at 18 to 36 inch centers (450mm-900mm) and reinforced nominally with a single #4 (#13) rebar. These holes are drilled to the base of the suspect material in order to create added arching and hold the cut face. (See Figures 5.17 and 5.18)

Vertical elements are also used to allow the upper row of soil nails to be depressed so that they can pass under near-surface utilities. Drilled holes, reinforced with pipe or small wide flanged sections, are used to create a larger cantilever than might normally be seen with conventional shotcrete applications (see Figures 5.19 and 5.20).

Finishes

Shotcrete is applied by the wet process. It is blown on under air pressure and stacked in layers from the bottom of a lift up (see Figure 5.21). Temporary shotcrete is placed and may or may not be struck off with a screed. As such, it has a very rough texture.

Permanent shotcrete walls can be finished and brought to a very clean surface. They are screeded flat and then finished with a wood float (see Figure 5.22). The surface approaches that of a cast-in-place wall for flatness and it has a slightly sanded texture. When desired, a textured finish can be applied to the shotcrete by imprinting. False joint lines or other relief features can also be trowelled into the finished face to give it a cast-in-place appearance (see Figure 5.23).

FIGURE 5.16 Drainage fabric and rebar placement prior to shotcrete application, Redmond, W.A. (*Courtesy of Condon-Johnson & Associates, Inc. Seattle, WA*)

FIGURE 5.17 Vertical elements for face stability, Vancouver, WA. *(Courtesy of Drill Tech Drilling and Shoring Inc. Antioch, CA)*

FIGURE 5.18 Vertical stability in this dessicated soil is enhanced with vertical elements at 3 feet (900 mm) c/c. Glendale, CA. (*Courtesy of Drill Tech Drilling and Shoring Inc. Antioch, CA*)

210 EARTH RETENTION SYSTEMS

FIGURE 5.19 Eight inch (200 mm) diameter vertical elements to incease cantilever, Seattle, WA. *(Courtesy of Golder Associates Inc. Redmond, WA)*

FIGURE 5.20 Vertical elements at 6 foot (1.83 m) centers cantilever the upper portion of this permanent wall and permit placement of the first lift of soil nails at 7 feet (2.13 m) below grade. Seattle, WA. *(Courtesy of Golder Associates Inc. Redmond, WA)*

FIGURE 5.21 Shotcrete application, Mukilteo, WA. *(Courtesy of Condon-Johnson & Associates, Inc. Seattle, WA)*

FIGURE 5.22 Shotcrete finishing, Redmond, WA. (*Courtesy of Condon-Johnson & Associates, Inc. Seattle, WA*)

214 EARTH RETENTION SYSTEMS

FIGURE 5.23 Shotcrete wall with detailed finish, Agura Hills, CA. *(Courtesy of Condon-Johnson & Associates, Inc. Los Angeles, CA)*

Shotcrete is applied in layers and often concrete of different batches will cure with a different color. This color variation must be expected and if color variation is deemed to be a visual problem, it can be overcome with a spray-applied solid body stain.

If visual appearance is paramount, shotcrete can be tooled to take on a natural rock look. The finished face can then be stained with various stains to complete the effect (see Figure 5.24).

5.3 Excavation and Backfill

Excavation adjacent to lagged or shotcrete shoring systems should always be performed by backhoe or tracked excavator. These machines cut a soil face by cutting and pulling away from the soil mass leaving a relatively undisturbed face. Usually the mass excavation of the site is made without cutting for the lagging. The mass can be cut with loaders, scrapers or backhoes. In the vicinity of any lagging, a berm is left which is later removed just prior to the lagging operation (see Figure 5.25). Depending on the stability of the cut soils, the final face of a

FIGURE 5.24 Stained textured walls, Felton, CA. *(Courtesy of Condon-Johnson & Associates, Inc. Oakland, CA)*

soil nail and shotcrete application can be cut either prior to or after installation of the soil nails. The final trimming of the soil face for lagging should be done in a fashion such that there is little or no gap left between the lagging plank and the soil face. Loaders excavate by pushing and lifting. This action tends to disturb materials in front of the excavation face as well as disturbing lagging or shotcrete placed in previous lifts.

While it may be possible to excavate a shored excavation in 10-12 foot (3.0-3.6 m) lifts matching the tieback or raker elevations, lagging cuts must be performed in lifts that can be safely exposed. In general, lifts of 4-5 feet (1.2-1.5 m) are preferred. It is still possible to mass excavate a 10-12 foot (3.0-3.6 m) cut while coordinating with the lagging operation. A cut of 4-5 feet (1.2-1.5 m) for lagging is made and then a berm is cut down to the base of the mass excavation lift desired (see Figure 5.26). A working surface sufficient to install the lagging must be left. Once the lagging lift is installed, the berm is removed and the lagging continued down to meet the mass excavation.

It is very important that any gap between the lagging and the soil face be filled. The material used can often be in situ cohesionless (sandy) materials. The practice of pumping CDF (controlled density fill) behind the lagging should not be instituted as a general solution as it creates areas which cannot drain and therefore may develop water build up. Solutions which require overexcavation of the soil face to permit placement of drainage fabric and free draining gravel are sometimes specified. These designs are not constructible in situations which require placement of lagging in multiple lifts as the backfill material falls out when the subsequent lifts are exposed.

FIGURE 5.25 Mass excavation ahead of lagging, note berm left against lagging face which is removed just prior to lagging installation, Seattle, WA. *(Courtesy of Condon-Johnson & Associates, Inc. Seattle, WA)*

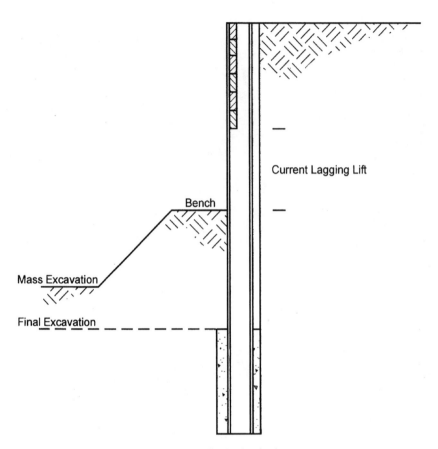

FIGURE 5.26 Typical excavation section indicating lagging berm.

CHAPTER 6
SHORING USES

When selecting the type of shoring for a particular project, it is important that a rational appraisal of its ultimate usage be made. The design intent can have a great influence on the type of shoring chosen and can affect even the engineering method and factors of safety used in its design. Once a shoring method is chosen, it is sometimes difficult to change to another system. Some systems, if selected, might be adaptable to a change of usage while others simply cannot be revised. A change in use might result in the abandonment of the initial shoring in order to construct a system compatible with the revised intent. The following are the types of uses and their constraints.

6.1 TEMPORARY

Temporary shoring systems are just that—temporary. This is not to say that they are flimsy or unsafe, but they are designed with the understanding that they will be in place and load bearing for a finite period of time. The period envisioned may be as short as a number of hours in the case of trench boxes (see Chapter 3.2) or as long as two years for deep building excavations. The *PTI Manual for Soil and Rock Anchors* specifies that any exposure longer than 24 months should be considered permanent, at least in terms of corrosion protection for the anchor components of the wall system. Exposures to particularly aggressive soil conditions may require corrosion protection for even shorter periods of time. See the bibliography in this text for references on the *PTI Manual*. Some typical temporary shoring systems are shown in Figures 6.1 through 6.3.

FIGURE 6.1 Conventional tiedback soldier pile and lagging wall, Seattle, WA. (*Courtesy of Condon-Johnson & Associates, Inc. Seattle, WA*)

FIGURE 6.2 Temporary soil nail wall, Portland, OR. *(Courtesy of Condon-Johnson & Associates, Inc. Seattle, WA)*

FIGURE 6.3 Temporary sheet pile wall, Everett, WA. (*Courtesy of Hurlen, Inc. Seattle, WA*)

Aside from trench boxes previously mentioned, the types of shoring considered to be temporary are sheet piling (Chapter 3.1), timber shoring (Chapter 3.3), lightweight shoring (Chapter 3.4), soldier pile and lagging (Chapter 3.5), soil nailing (Chapter 3.6) and secant walls (Chapter 3.7)

Shoring applications, such as trench shoring or the shoring of an excavation for the installation of a tank, or larger excavations for building basements, are included in this category. Similarly, excavations for bridge abutment construction or retaining wall construction are appropriately considered temporary.

Temporary excavation support is usually designed based on active soil parameters. (K_a; see Chapter 8.4). The exception to this statement is the case where adjacent buildings or utilities are so sensitive that the types of movements generally experienced in allowing the retained soils to develop an active state of stress would permit too much settlement. In these cases, at-rest analyses (K_o; see Chapter 8.6) are used.

It is not customary to design temporary excavation support for seismic loading. This is not to say that the designer takes a cavalier approach and is betting that a seismic event will not occur. Experience has shown that temporary excavation support methods tend to be flexible enough that moderate seismic events do little or no damage to these systems. These observations have been gathered from Loma Prieta in 1989—7.1 on the Richter Scale, Northridge in 1994—6.7 on the Richter Scale, and Nisqually 2001—6.8 on the Richter Scale.

Temporary shoring systems are designed to provide no long-term support for either the soil mass or the structure constructed adjacent to them. In some cases, such as sheet piling, soldier pile and lagging, soil nailing and secant walls, some or all of the system may be left in place and abandoned. Regardless of whether it is taken out or left in place, the shoring system is, by definition, assumed to have no structural value once the permanent structure is in place and the excavation backfilled.

6.2 PERMANENT

In cases where the engineer has decided to permanently retain the earth with a shoring system, long-term design principles are used. The types of systems suitable for permanent applications include sheet piling (Chapter 3.1), soldier pile and lagging (Chapter 3.5), soil nailing (Chapter 3.6), secant walls (Chapter 3.7), cylinder walls (Chapter 3.8), slurry walls (Section 3.9), micropiles (Chapter 3.10), and underpinning (Chapter 3.11). Many conventional retaining wall situations can be economically dealt with using shoring methods outlined in Chapter 3 (see Figure 6.4). These methods can also be utilized to repair existing failed retaining walls (see Figure 6.5).

Permanent installations are almost always designed using at-rest principles (K_o; see Chapter 8.6). This is because a number of events conspire to increase loading on permanent earth retaining structures, such as freeze/thaw cycles, wet/dry cycles, deterioration of drainage systems, strain softening, and creep (see discussion in Chapter 14).

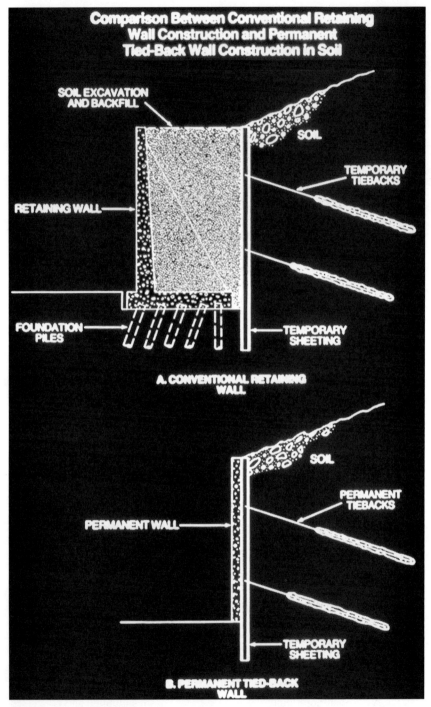

FIGURE 6.4 Permanent soldier pile and lagging with fascia treatment in lieu of conventional cast-in-place retaining wall. *(Courtesy of Schnabel Foundation Co., Inc. Houston, TX)*

SHORING USES 225

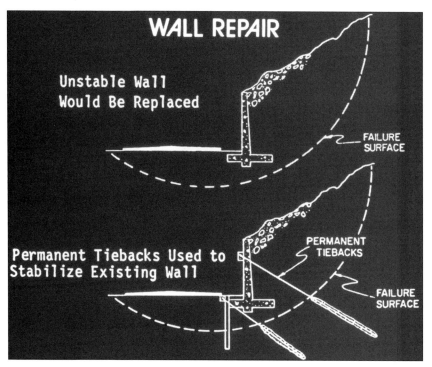

FIGURE 6.5 Repair of failed cast-in-place retaining wall utilizing tiebacks-schematic. *(Courtesy of Schnabel Foundation Co., Inc. Houston, TX)*

Permanent installations should always be designed with proper accommodation for seismic occurrences. Samples of permanent walls are shown in Figures 6.6 through 6.8.

Corrosion protection must be addressed in permanent installations. See recommendations of the *PTI Manual for Soil and Rock Anchors*. Adequate drainage must also be addressed to prevent unwanted hydrostatic buildups or leakage.

There can be a number of reasons to use one of these earth retaining structures as a permanent system. In side hill cuts, they can be a very economic form of permanent retaining wall. In some building construction cases, it is convenient to take lateral loads out of an earth cut and not force the building frame to handle these loads. This is particularly apropos in cases where building basements are set in side hill cuts with the retained earth being much higher on one side of the building than the other (see Figure 6.9).

FIGURE 6.6 Permanent soldier pile and lagging wall, Alamo, CA. *(Courtesy of Condon-Johnson & Associates, Inc. Oakland, CA)*

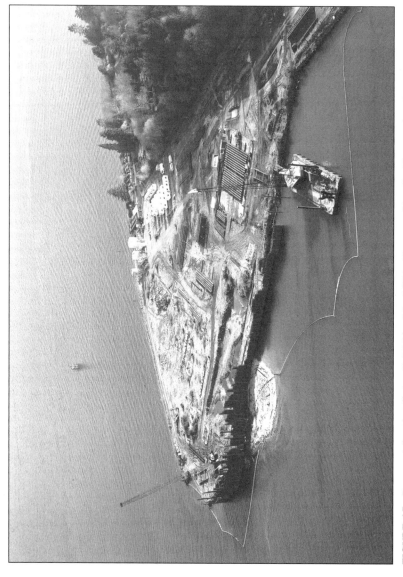

FIGURE 6.7 Permanent sheet installation, Bainbridge Island, WA. (*Courtesy of Aerolist Photographers Inc. Renton, WA*)

FIGURE 6.8 Permanent soil nailing, Mukilteo, WA. (*Courtesy of Condon-Johnson & Associates, Inc. Seattle, WA*)

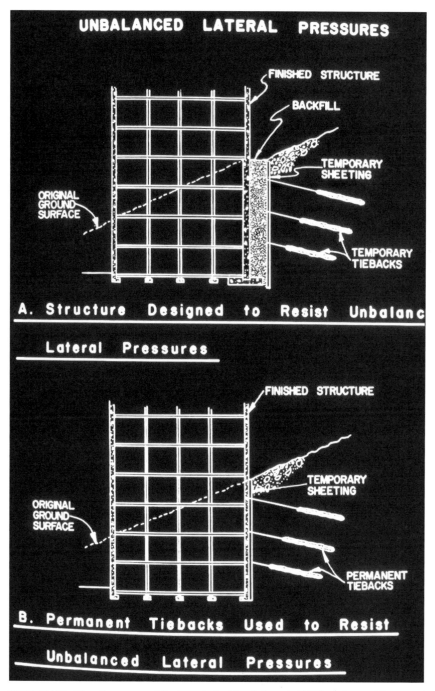

FIGURE 6.9 Use of permanent tiebacks to deal with unbalanced sidehill cut forces. *(Courtesy of Schnabel Foundation Co., Inc. Houston, TX)*

6.3 TEMPORARY/PERMANENT—THE HYBRID APPLICATIONS

In recent years there has been an increased use of hybrid applications of temporary and permanent shoring. These applications require attention to detail. The temporary portions of the work can be designed as such, while the permanent must be designed to permanent standards. This can sometimes be quite confusing to analyse because it is possible to have construction loads (the temporary application) that are higher than permanent loads.

6.3.1 Temporary Soldier Pile and Lagging with Permanent Tiebacks

In cases where the permanent structure requires tiebacks, it is sometimes possible to design the temporary shoring to utilize the permanent tiebacks for lateral support. Temporary soldier piles and lagging are constructed together with the permanent tiebacks (see Figure 6.10). When the permanent structure is constructed inside the shored excavation, it is attached to the tiebacks through load transfer devices such as studded plates. In these cases, the external walls, which are connected to the tiebacks, will be subjected to the entire lateral earth pressure but the internal building diaphragm is spared the lateral load.

6.3.2 Permanent Soldier Piles with Temporary Shoring System.

In cases where the soldier piles are to be used to provide vertical stiffening of the permanent wall system, it is possible to design the soldier piles to perform both the task of temporary earth support and then marry the piles to the permanent structure to stiffen the walls. This can be done by placing studs on the soldier piles which are then included in the concrete wall pour.

6.3.3 Soil Nailing—Temporary Nails, Permanent Fascia

The shotcrete fascia of a soil nail system can be designed to be the permanent wall of a finished structure. If the permanent structure is designed to support the lateral load of the soil through its flooring system, then the soil nails are designed as temporary. Of course, the design of the permanent basement wall with its code concerns for cover and steel minimums is much different than that of a temporary soil nail wall (Chapter 4.7). In these cases, the resultant wall is much thicker (8 inch (200 mm) minimum) than a temporary fascia (4 inch (100 mm) nominal), (see Figure 6.11). There is additional discussion of this in Chapter 6.4.

FIGURE 6.10 Temporary shoring utilizing permanent tiebacks to support soldier pile and lagging wall. Permanent tiebacks were used to support the back wall of the structure against a potential slide mass, Seattle, WA. (*Courtesy of Condon-Johnson & Associates, Inc. Seattle, WA*)

FIGURE 6.11 Temporary soil nails supported this permanent fascia wall until the remainder of the substructure was completed, Boise, ID. (*Courtesy of Condon-Johnson & Associates, Inc. Seattle, WA*)

6.3.4 Soil Nailing—Permanent Nails, Temporary Fascia

In a slightly different case than permanent tiebacks, it is possible to design a building substructure such that the basement walls are spared most of the load imposed by lateral earth pressures, while the internal diaphragm of the building does not experience any loading from the soils. In this case, the soil nailing is designed as a permanent application and connections are detailed to attach the soil nails to the permanent walls. The shotcrete fascia which forms the facing during excavation and provides protection against sloughing is designed as a temporary wall.

6.4 TOP-DOWN FASCIA CONSTRUCTION

The use of this term should not be confused with a system practiced more often in Europe where the permanent building substructure is constructed as a whole from the ground down. On some European jobs, floors and walls are cast as the excavation progresses and the excavation is performed by digging below the cast in place concrete and removing the soil through the completed structure. Top down fascia construction is the process by which the permanent fascia wall of either a soldier pile system or a soil nail system is built as the excavation progresses. Examples of top down fascia systems are shown in Figures 6.12 through 6.16. These installations are almost always done by shotcrete methods. There are several advantages and some disadvantages to these systems which should be seriously considered prior to deciding to adopt this method.

6.4.1 Advantages

- By eliminating temporary lagging in soldier pile and lagging or temporary shotcrete fascia in soil nailing, significant cost savings can be achieved.
- By constructing the permanent wall as the excavation progresses, it is possible to save considerable time in the construction schedule. When the excavation is complete, the external basement walls are already constructed, removing that operation from the critical path on the schedule.

6.4.2 Disadvantages

- Placing the wall in shotcrete lifts necessitates more reinforcing steel splices.
- Drainage and waterproofing are far more difficult to perfect in a top down installation and it is virtually impossible to install a wall that does not have the potential to leak, or at least effervesce.
- The weight of a permanent wall (which may be as thick as 24 inches (610 mm) depending on the particular application) may be too heavy for the soil nail system to carry during construction without the addition of strut nails (Chapter 4.7). This added cost must be considered.

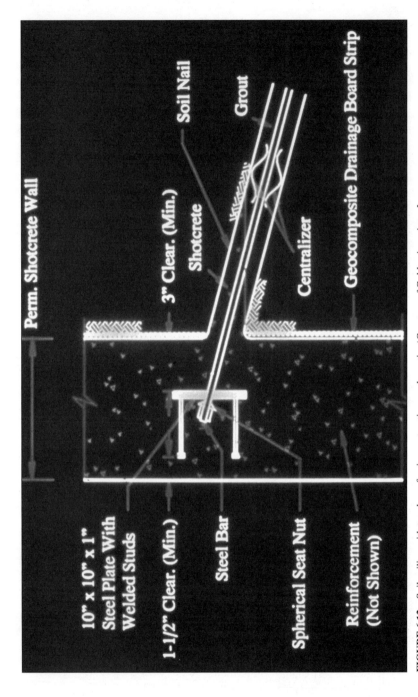

FIGURE 6.12 Soil nailing with top down fascia construction—schematic. (*Courtesy of Golder Associates Inc. Redmond, WA*)

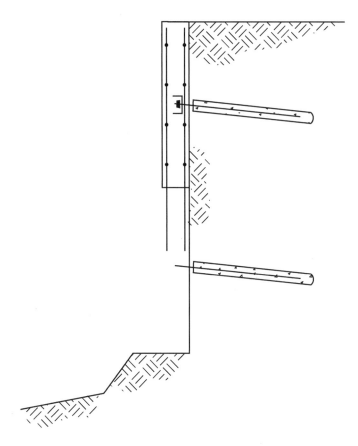

FIGURE 6.13 Top down construction—excavation schematic.

- The integration of the various trades involved in excavation, reinforcing steel placement, waterproofing application, shotcreting, and concrete finishing is very difficult, especially on a small site. Continuity of work for all concerned is very difficult to achieve.

6.5 Slide Control—Repair

Some shoring systems are very applicable to prevent, control or repair damages from land sliding. These systems include sheet piling (see Chapter 3.1), soldier pile and lagging (Chapter 3.5), soil nailing (Chapter 3.6), secant walls (Chapter 3.7), cylinder walls (Chapter 3.8), and slurry walls (Chapter 3.9).

FIGURE 6.14 Soil nailing with top down fascia construction, Redmond, WA. (*Courtesy of Condon-Johnson & Associates, Inc. Seattle, WA*)

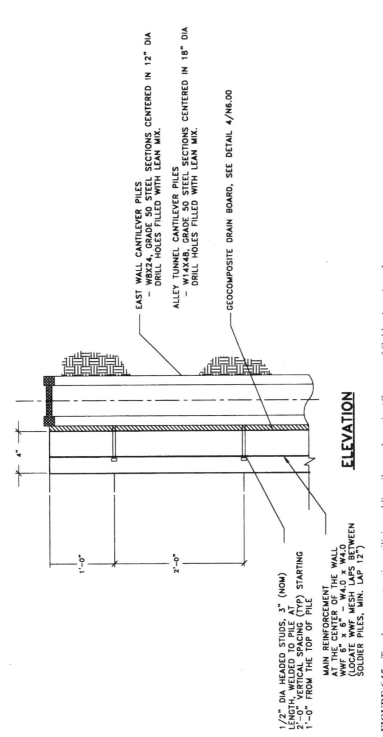

FIGURE 6.15 Top down construction utilizing soldier piles—schematic. (*Courtesy of Golder Associates Inc. Redmond, WA*)

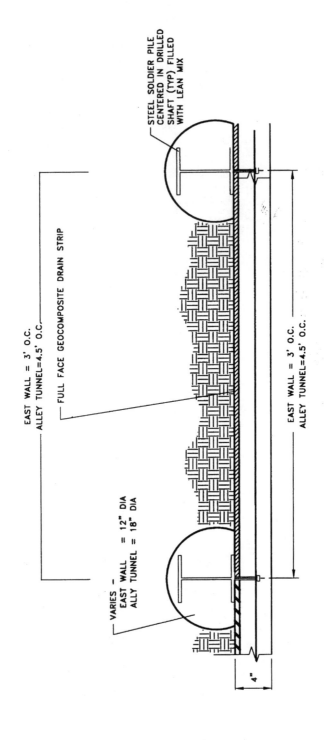

FIGURE 6.15 (*continued*) Top down construction utilizing soldier piles—schematic. (*Courtesy of Golder Associates Inc. Redmond, WA*)

FIGURE 6.16 Soldier piling with top down fascia construction, Kirkland, WA. (*Courtesy of Condon-Johnson & Associates, Inc. Seattle, WA*)

The design of a shoring system in a sliding situation is handled just like a permanent wall (Section 6.2) with the following additions. Often a slide plane will pass below the level of the base of the downhill excavation. In this case, the vertical element (be it a sheet pile or a soldier pile) must be designed to intercept and strengthen the slide plane so as not to endanger the wall in the manner discussed in Section 10.5. A typical slide repair project is shown in Figures 6.17 and 6.18.

In addition, it is often desirable to provide debris flow constraint capabilities to the retention wall. This is done by increasing the height of the wall some distance above the height of the up-hill finished grade. A typical debris wall is shown in Figures 6.19 through 6.22. In order to do this, the wall must be designed to handle not only the potential load of debris retained by the "catchment" wall, but also the dynamic forces involved in downhill movement of the debris being caught.

Soil nailing can be used as a dowelling process to enhance the shear strength of *in situ* soils to prevent sliding. Figure 6.23 is a photo of a soil nail project designed to strengthen the excavated slope on a dam abutment.

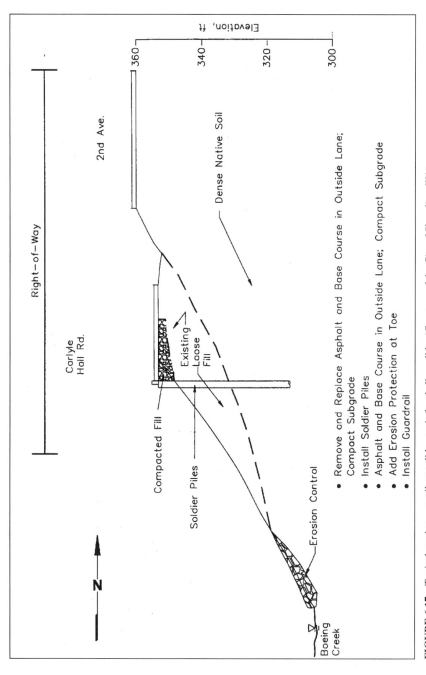

FIGURE 6.17 Typical section—cantilever slide repair for shallow slide. (*Courtesy of the City of Shoreline, WA*)

FIGURE 6.18 Tangent pile wall with permanent tiebacks supported the excavation against an ancient slide mass, Seattle, WA. *(Courtesy of Condon-Johnson & Associates, Inc. Seattle, WA)*

SHORING USES 243

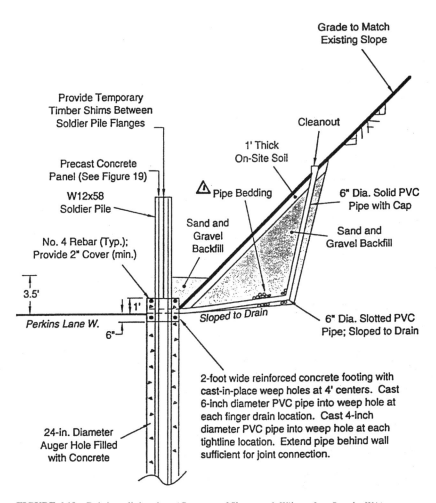

FIGURE 6.19 Debris wall drawing. *(Courtesy of Shannon & Wilson, Inc. Seattle, WA)*

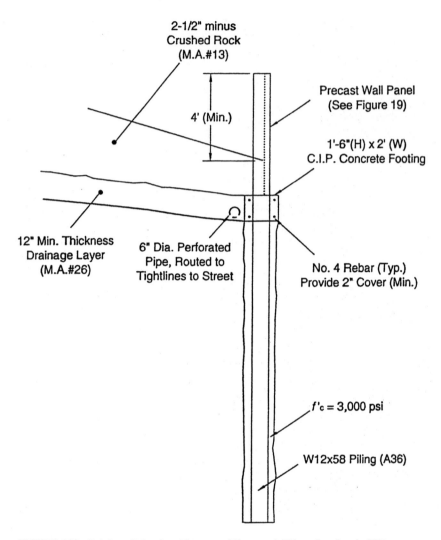

FIGURE 6.20 Debris wall drawing. *(Courtesy of Shannon & Wilson, Inc. Seattle, WA)*

FIGURE 6.21 Soldier pile and precast lagging form a debris wall, Seattle, WA. (*Courtesy of Condon-Johnson & Associates, Inc. Seattle, WA*)

FIGURE 6.22 Debris wall from back side indicating catchment area for moving slide debris. (*Courtesy of Condon-Johnson & Associates, Inc. Seattle, WA*)

FIGURE 6.23 Use of soil nailing to reinforce a dam abutment, Salmon, ID. (*Courtesy of Condon-Johnson & Associates, Inc. Seattle, WA*)

CHAPTER 7
INVESTIGATIONS

A number of investigations must be undertaken prior to the design and construction of an excavation. These include the drilling of soil borings and developing of Geotechnical Reports, the performing of Pre-Construction Surveys of adjacent properties and utilities, and the analyzing of the location of those utilities to verify potential interferences.

7.1 GEOTECHNICAL REPORTS

Prior to the design and installation of any underground improvement, an analysis of subgrade materials and conditions must be performed. This is usually done by drilling test borings and performing lab and field tests to characterize the subsurface conditions and ascertain various parameters of the soils. While borings are not always necessary, some form of subsurface investigation must be performed. The reliance on shallow test pits should only be undertaken when local knowledge of the area can supplement and confirm the findings of the test pits.

Soil borings are taken to develop an understanding of strength parameters of the soils and ground water conditions and document any incidence of contamination which may be evident. Borings can be undertaken in a number of ways to produce meaningful information. This chapter will not delve into the many and varied methods of drilling soil borings, but will concentrate on the useful data which should be available from those borings.

A geotechnical report which is helpful to both the designer and constructor of a shored excavation will include the following materials:

- A listing of all soils encountered, described in accordance with a well recognized soil classification system such as the Uniform Soil Classification System. All soils encountered should be carefully logged so as to present an accurate cross section of the soils encountered at the boring location.
- A description of any water levels encountered in the boring, both at the time of drilling and some time later when static levels can be determined.
- A discussion of the applicability of the boring results to known geologic mapping of the area.
- A description of any evidence of hazardous or contaminated materials noted in the borings.
- A full disclosure of any in-hole testing which was undertaken together with the test results such as standard penetration tests (SPT), cone penetration tests (CPT), shear vanes, and slugging tests. If samples were retrieved, the location where they can be viewed should be indicated.
- An accurate plan indicating boring locations referenced to property lines, building lines, or easily identified monuments. Elevation of the top of all borings should be referenced to an easily identified datum, preferably geodetic.
- A discussion of the relevant parameters of the soils investigated. This discussion should include parameters such as c, ϕ unit weight (γ), moisture content, grain size analysis and incidence of boulders and cobbles. In rock, such parameters as Rock Quality Designation (RQD), compressive strength, and the incidence of fissuring should be provided. If available, strike and dip data is also helpful.
- A discussion of the water levels. Is the water indicated perched, or is it representative of the true water table? If dewatering is indicated, what is the anticipated hydraulic conductivity?
- Although it is preferred that the Geotechnical Report give c and ϕ values, if they are not available, a discussion of suggested apparent earth pressures or earth pressure coefficients is necessary.

It is recognized that this information is probably not all reported in the preliminary Geotechnical Report, but it must be determined prior to the design and installation of a safe shored excavation.

7.2 PRE-CONSTRUCTION SURVEYS

Pre-construction surveys are carried out just as the phrase says—prior to construction. Some of the information gleaned from these surveys is necessary for design of the shoring system, while the remainder is an important log of the pre-existing condition of the adjacent property. These surveys are done for the dual purpose of protecting the owner, engineer and contractor on a project against liability for pre-existing conditions, as well as determining the type of shoring required to ensure that no damage occurs to the adjacent property from the planned excavation procedures.

Required pre-design information regarding adjacent properties includes:

- Description of the size, shape and type of construction of any adjacent facilities such as buildings, roads, retaining walls and utilities. Report on any as-built documentation of these adjacent facilities
- Accurate location and depth of adjacent building basements and footings, together with an estimate of the location and depth of buried utilities.
- An estimate of the loads on adjacent building footings together with a disclosure of the competence of the structure.
- A clear definition of any easements, fire lanes, or building exits adjacent to the proposed excavation.

Pre-Construction information which should be collected prior to starting construction includes:

- Crack surveys including pictures, or videos to record pre-existing conditions. Figures 7.1 through 7.5 are typical pre-construction photos which define the existing condition of the adjacent properties
- Detailed water level and water quality sampling of wells in the vicinity.
- Traffic surveys if excavation is likely to affect neighboring businesses.

7.3 UTILITY LOCATES

Although normally considered to be part of the pre-construction survey, this issue has been given its own segment due to the importance of due diligence in this matter. While overhead utilities can be seen and are often moved prior to construction, buried utilities, which are far greater in number, must be located prior to construction. Not only does the striking and subsequent disruption of service represent a significant potential liability to the project participants, it can be also be a considerable safety hazard to the personnel directly involved.

While utility locates are attempted by the engineer prior to the design of the shoring, this locate information should never be relied upon. The construction team (owner, general contractor, and specialty contractor) should undertake a utility location survey prior to construction. This survey should include:

- A review of all available as-built drawings
- Utility marking by a coordinated utility marking service. See Figure 7.6 for a typical marking by a one-call locating agency.
- Utility locates by each individual utility
- Opening of manholes and vaults to measure the depth and location of inverts of adjacent utilities
- Pot holing as necessary to locate utilities.

FIGURE 7.1 Pre-construction photos depicting overall structure. *(Courtesy of Murray Franklin Inc. Redmond, WA)*

FIGURE 7.2 Pre-construction detail of terra cotta jointing. *(Courtesy of Murray Franklin Inc. Redmond, WA)*

FIGURE 7.3 Pre-construction photos of interior conditions. *(Courtesy of Murray Franklin, Inc. Redmond, WA)*

FIGURE 7.4 Pre-construction photos of known defects. *(Courtesy of Murray Franklin Inc. Redmond, WA)*

FIGURE 7.5 Pre-construction photos of surrounding sidewalks. (*Courtesy of Condon-Johnson & Associates, Inc. Seattle, WA*)

FIGURE 7.6 Pre-construction locates of utilities marked on sidewalk. *(Courtesy of Condon-Johnson & Associates, Inc. Seattle, WA)*

As you will notice, each of the steps becomes increasing more difficult and invasive for the project team. However, the team should continue to fulfill each of these steps until such time as the utilities are located with a high degree of certainty.

Aside from the disruption to the utility caused by their breakage, utility damage can have a very detrimental effect to the project. Drilling into gas or electric lines can risk fatal injuries for any nearby personnel. Drilling and grouting operations that damage and subsequently fill sewers or waterlines can be very expensive to fix. Damage to any adjacent utility can cause severe disruption to the project schedule while the utility is repaired.

Successful excavation can only be performed when a complete understanding of the soils and adjacent facilities exists. Only by performing the above-mentioned tests in a diligent and thorough manner can the project team assure themselves that they have achieved this understanding.

CHAPTER 8
ENGINEERING PROPERTIES OF RETAINED SOILS

In order to perform a shoring design, it is necessary to use several engineering parameters and understand some basic concepts. Some of the parameters are measured, some developed and others are derived. Together they form the input data for the empirical and analytical methods used to design the shoring systems used today. The empirical design methods are based on the systematic collection and analysis of data obtained over many years on earth retention systems such as braced and tied-back soldier pile walls. For other systems, like soil nailed walls, analytical methods are used. Both these approaches accurately and effectively predict lateral earth pressures.

Because of layering of dissimilar soils, not all earth retention systems lend themselves to analytical methods. Lateral earth pressures can also be influenced by the types of restraint used. This is called soil-structure interaction and further confounds the solution of earth retention problems by the use of first principles. The inputs necessary for these design methods include the following.

8.1 ANGLE OF INTERNAL FRICTION

The angle of internal friction defines the increase in shear strength of a soil with increasing confining pressure. It is calculated by plotting a series of triaxial tests as Mohr Circles on a plot of principle stress vs. shear strength. This plot is called a Mohr circle diagram (see Figure 8.1). The asymptote of several Mohr circles is called the Mohr-Coulomb envelope. The angle of internal friction is the slope of the Mohr-Coulomb envelope and is defined in degrees from the horizontal. It is most pronounced in cohesionless soils (sands and gravels) and approaches zero in soft cohesive soils such as soft clay. The angle of internal friction can be also be determined by cone penetration tests, or laboratory tests completed on undisturbed samples taken in the field. The angle of internal friction, phi, (ϕ) can be estimated from standard penetration tests. A table of this correlation is shown as Table 8.1.

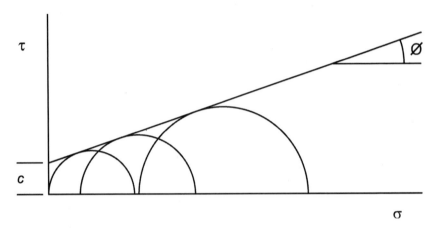

FIGURE 8.1 Mohr circle diagram.

TABLE 8.1 SPT vs. phi.

Type of Soil	Penetration Resistance N	Angle of Internal Friction	
		Peck (1974)	Meyerhof (1956)
Very Loose Sand	less than 4	less than 29	less than 30
Loose Sand	4-10	29-30	30-35
Medium Sand	10-30	30-36	35-40
Dense Sand	30-50	36-41	40-45
Very Dense Sand	over 50	greater than 41	greater than 45

Table 8.1 From Winterkorn & Fang

8.2 COHESION

Cohesion is a property exhibited in fine-grained soils (clays and silts), which is the result of atomic attractive forces between soil particles. These forces allow the material to exhibit shear strength, even when no confining pressure is available. Cohesion is the intercept of the Mohr-Coulomb envelope with the shear strength axis where the principle stress equals zero. Figure 8.1 illustrates a typical Mohr Coulomb envelope of a cohesive soil. In soft clays where ϕ is negligible, the cohesion approaches the measured shear strength of the *in situ* soils. In cases where triaxial tests are not available, the cohesion can be developed using cone penetration tests. It can also be approximated by using one half of the unconfined compressive strength (UU) as derived from pocket penetrometer tests. Cohesion, c, is a strength parameter and is expressed in units of pressure.

8.2.1 Apparent Cohesion

Some cohesionless soils will exhibit characteristics of cohesive soils in that they will stand vertically when cut. The reason these sands and gravels can do so is because of their moisture content. Some cohesionless soils, with moisture contents that are dry of saturation, will have their particles bound together by capillary attractive forces. These forces in the water molecules hold wetted soil particles together to form a weak cohesion. The phenomenon is called apparent cohesion. Apparent cohesion can also be the result of particle cementation caused by mineralogy or thixotropic action. Thixotropic action is the result of previous high stress history.

In the earth retention field, apparent cohesion will permit a vertical face of an excavated cohesionless soil to stand for at least a short period of time. This occurrence is called stand-up time and its existence is absolutely critical for lagging and soil nailing. In cases where the apparent cohesion is the result of capillary action, apparent cohesion may disappear with time as the exposed soil dries. Because of its lack of permanence, this apparent cohesion is not a property which is relied on for any calculation of soil strength.

In cases where the apparent cohesion is the result of mineralogy or thixotropic action, it is quantified by the methods detailed for defining cohesion (Chapter 8.2).

8.3 UNIT WEIGHT OF SOIL.

The wet weight (soil and entrained moisture) of a specific volume of soil is known as its unit weight. It is commonly referred to as γ_s or γ (gamma) and is expressed as weight per unit volume.

8.4 ACTIVE PRESSURE

The theory of earth pressure, as advanced by Rankine, defined a wedge of soil which would move if not restrained. The Rankine wedge is outlined in Figure 8.2. It is most easily demonstrated in the case where cohesion is equal to zero and we deal with the angle of internal friction only. Rankine held that, when a face was cut in soil, a wedge defined by an angle measured from the vertical axis equal to 45^0-$\phi/2$ from the toe of the excavation would be caused by gravity to try to move downward and outward. This gravitational force would be counteracted by the shear stresses acting on the line AB which defines the back of the wedge (the active wedge). The unbalanced force or resultant of these two forces is a function of the weight of the soil.

The function is known as the coefficient of active earth pressure and is designated as K_a. It is derived as

$$K_a = \tan^2(45- \phi/2) \qquad (8.1)$$

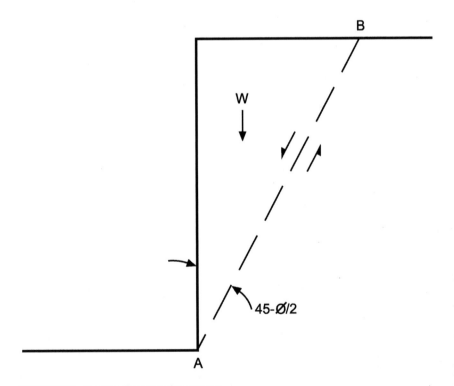

FIGURE 8.2 Rankine diagram-active pressure.

ENGINEERING PROPERTIES OF RETAINED SOILS 263

8.5 PASSIVE EARTH PRESSURE

Rankine held that if a force were externally applied to a face of soil in an attempt to force it back into itself, the force at equilibrium would be the sum of the gravitational load of the failure wedge plus the summation of the shear stress on the wedge plane defined as CD (see Figure 8.3). The angle defining the failure wedge is defined as $45^0 + \phi/2$ from the vertical axis. The force required to move this wedge of soil is a function of the weight of the soil.

The function is known as the coefficient of passive earth pressure and is designated as K_p. It is derived as

$$K_p = \tan^2(45+\phi/2) \quad (8.2)$$

8.6 AT-REST PRESSURE

To hold an excavated face of soil in place without the use of the shear strength of the soil is known as the at-rest condition. The at-rest pressure is a function of the weight of the soil. The factor defining that function is known as the coefficient of at-rest earth pressure and is designated as K_0. It is approximated as

$$K_0 = 1 - \sin \phi \quad (8.3)$$

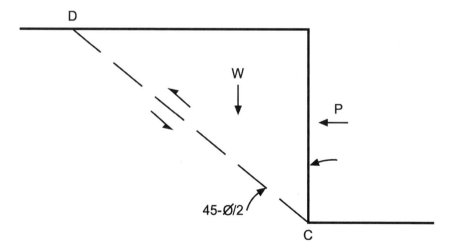

FIGURE 8.3 Rankine diagram-passive pressure.

8.7 HYDROSTATIC PRESSURE

Hydrostatic pressure is the force that is exerted on a shoring system by water that is retained behind the shoring system. As water is known to weigh 62.4 pcf (1 T/m^3) and defined by the symbol γ_w, the pressure acting at any point on a shoring system which impounds water behind it is equal to the depth of the water X 62.4 pcf (1 T/m^3). This impoundment will also have the effect of reducing the unit weight of soil to the buoyant weight (γ')

$$\gamma' = \gamma_s - \gamma_w \qquad (8.4)$$

8.8 ARCHING

Arching is that phenomenon in a soil which permits it to transfer load to points of rigidity similar to the way in which an arch bridge shifts its weight to its piers and abutments (see Figure 8.4). This phenomenon allows even cohesionless soils to stand temporarily between points of rigidity when unsupported, and sometimes allows the designer to reduce the design stresses acting on parts of the shoring system. Arching acts not only in the horizontal plane of the shoring wall, but also in the vertical (see Figure 8.5). It is most evident in cohesionless soils (sands and gravels) and approaches zero in soft, fine grained soils (clays).

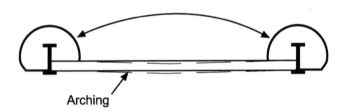

Arching

Deflection Required For
Development of Soil Arching

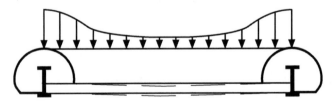

Result ant Earth Pressure

FIGURE 8.4 Horizontal arching—note that the uniformly distributed load is redistributed and reduced in the center of the span where deflection is greatest.

ENGINEERING PROPERTIES OF RETAINED SOILS

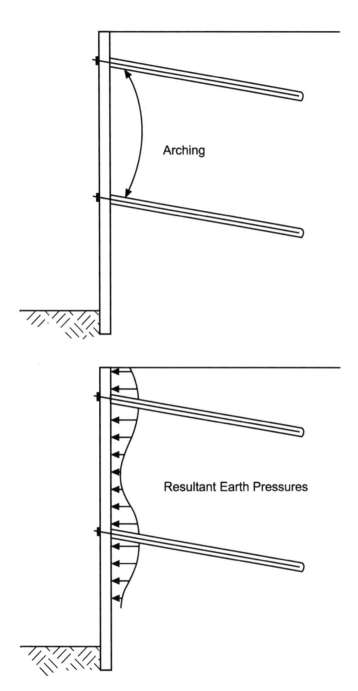

FIGURE 8.5 Vertical arching. As deflection occurs between points of stiffness, load is reduced in mid span and increased at nodes of stiffness.

These few concepts, when understood, will permit an engineer to comprehend the commonly used calculations used in the design of walls and to reconcile the forces acting on those walls. Chapters 9 and 11 will develop the use of these concepts in the design methods used for shoring.

CHAPTER 9
FORCES ON WALLS

Over the years, theorists have used many methods to analyze retaining walls and their effect on the adjacent soil mass, or to analyze the soil mass and its effect on the retaining wall. Today, almost all wall designs are based on one of three methods of analysis. These are:

- Earth pressure theory advanced by Coulomb and Rankine
- Apparent earth pressure advanced by Terzaghi and Peck
- Limit equilibrium developed from analysis work on the stability of earth slopes

Currently, many designers work seamlessly between these three theories, changing from one to another in mid-analysis almost without acknowledgement. As a result, a body of work exists to design walls which is largely based on experience and relies on successful previous case histories.

Other than to acquaint the reader with some of the differences in the various methods, no attempt will be made here to rigorously explain their intricacies. The bibliography of this text offers reading which can provide the reader with added information on these design methods.

Many designers have experienced difficulties with walls they have designed because they have simply focused on the wall as a lateral load resisting element. For the designer to focus only on the lateral loading indicated in these various design methods without examining the entire loading regime which may exist in a wall can lead to problems if not outright failures.

In order to design a shored wall, all the forces acting on the wall must be understood. Generally, the designer is primarily concerned with horizontal forces acting to topple the wall into the excavation. At times however, vertical loads on shoring must be considered. Also of great importance in soft soils (primarily cohesive) is the tendency for the base of the excavation to heave or exhibit instability. This tendency will greatly influence the design of a wall and must be considered in order to accurately understand the loads acting on the wall.

This chapter will outline the forces used to design shored walls of excavations. The lateral earth pressures are generally approximated by various diagrams which have been developed from years of observation and measurement. While the models used have been shown to be effective at predicting the lateral loads, in most cases they are empirical and are not based on any rigorous analysis from first principles.

The variability of soil stratigraphy, an imperfect understanding of the relationship between movement of the soil mass and the strength of its constituent soil layers, and the complexity of soil/structure interaction as it relates to the built facility has rendered attempts to create rigorous models of earth pressure and deformation difficult and implementation of such models virtually impossible.

The diagrams shown in this chapter represent those which are generally accepted for use today. Other diagrams or theories of earth pressure do exist and by their exclusion the author does not intend that they should be disregarded. This chapter will merely provide an understanding of how most of the shoring in use today is designed.

9.1 CANTILEVER SHORING

Cantilever shoring is most often created by using sheet piles (Chapter 3.1), soldier pile and lagging (Chapter 3.5), secant piles (Chapter 3.7), cylinder piles (Chapter 3.8), or slurry walls (Chapter 3.9). Cantilever shoring is the one case where it has been found that the Rankine model (see Chapter 8.4 and 8.5) of earth pressure theory will reasonably accurately predict forces on a wall. In cantilever shored situations, a wall accepts a horizontal force against it and resists the force by the rigidity of its embedment into the soils beneath the excavation. The embedded portion of the wall will develop a point of rotation, and passive forces will act on both sides of this point. This is called a moment couple (see Figure 9.1). The couple is the result of passive pressures acting on opposite sides of the wall embedment.

In cohesionless soils, the horizontal pressure acting on the walls and attempting to overturn them is directly proportional to the overburden pressure acting at that depth plus any surcharges which may be imposed on the ground surface (see Chapter 9.6). This vertical pressure is modified by a factor, K_a, (see Chapter 8.4) to define the horizontal pressure.

As you can see by the pressure diagram (Figure 9.2) the pressure is triangular. The figure is representative of the behavior of a granular soil where K_a is a function of internal friction angle Φ as discussed in Chapter 8.1 and 8.4. In Figure 9.3, a rectangular addition represents the surcharge loading of a generalized Uniformly Distributed Load (UDL) at the ground surface. Chapter 9.6 will discuss other types of surcharge loading and their effect on lateral pressures.

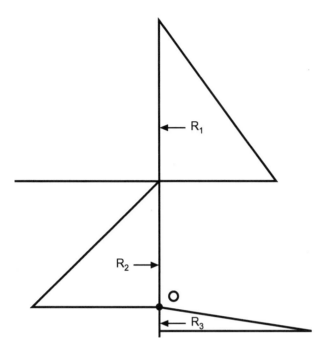

FIGURE 9.1 Cantilever force diagram.

The basic equation which defines lateral earth pressure at any point on the wall can be shown to be

$$P = K_a (\gamma H + q) \tag{9.1}$$

where
- P is the pressure at any point
- K_a is the coefficient of active earth pressure
- γ is the unit weight of the soil being retained (in the case of soil below the static water table it is defined as γ').
- H is the height of earth retained at the point of calculation
- q is the vertical component of the surcharge load at the depth considered

Although the earth pressure is triangular, in the case of cohesive soils the pressure diagram in its theoretical development is laterally shifted so that the upper portion of the diagram actually indicates a negative lateral pressure (see Figure 9.4). When you think about this it actually makes sense. A soil mass that has cohesion will stand vertically for some height. In this height, the pressure diagram indicates that no lateral restraint is necessary.

270 EARTH RETENTION SYSTEMS

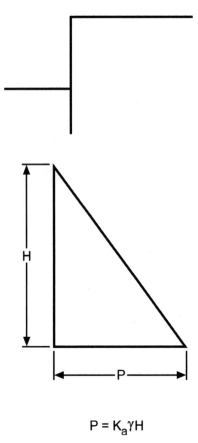

$$P = K_a \gamma H$$

FIGURE 9.2 Active pressure in cohesionless materials—triangular.

In cohesive soils, the lateral earth pressure from the soil mass is defined as

$$P = \gamma H - 2c \qquad (9.2)$$

where

P is the pressure at any point
γ is the unit weight of the soil being retained
H is the height of earth retained at the point of calculation
c is cohesion (see Chapter 8.2)

Although we acknowledge that the effect of cohesion in clayey soil indicates that there is no horizontal force in the upper levels of the cut, the reality of the performance of such a cut is that, over time, soils will dry and probably slake. This creates a dangerous situation for anyone below such a cut so that, if the cut

FORCES ON WALLS 271

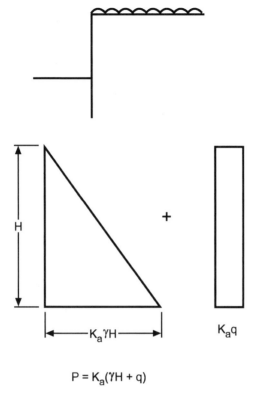

FIGURE 9.3 Active pressure in cohesionless materials—triangular plus surcharge.

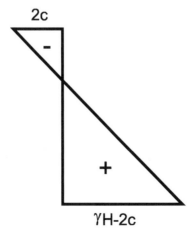

FIGURE 9.4 Active pressure in cohesive materials—triangular.

is higher than about four feet, the full face of the cut really must be shored. To design a cantilever wall in a cohesive soil, designers will rearrange the earth pressure diagram so that it begins with an intercept at zero at the ground surface and ends up at depth with the equivalent total load as indicated by Equation 9.2. This creates a kind of artificial coefficient of lateral earth pressure which can then be used to calculate the pressure at any point on the wall and permit the inclusion of lateral pressure due to surcharge.

Example 9.1 indicates a calculation method used to achieve an apparent lateral pressure coefficient in a cohesive soil.

EXAMPLE 9.1 CALCULATE EFFECTIVE K_a FOR SOFT CLAY
If c = 300 psf
 H = 15 feet
 γ = 120 pcf
P at point zero is equal to 120 (0) – (2) 300 = -600 psf
At Point 15, P is equal to 120 (15) – (2) 300 = 1200 psf
Total Load P_t = (1200 + (-600))•15/2 = 4500 plf
Set P at 0 equal to zero
Coefficient of active pressure can be calculated as follows
 P_t = ½ ($K_a \gamma H^2$)
 K_a = 2 Pt/γH^2 = 2•4500/120•225 = 0.33

Having developed this coefficient, it is now possible to design the cantilever wall in clays using Equation 9.1.

If a surcharge loading is applied at the top of the wall, it will be reflected as a horizontal pressure on the wall in the same manner as Figure 9.3.

In addition, if the wall is designed to retain water behind it, such as a sheet pile wall, slurry wall, or secant wall, without the use of relief drains, a further load which represents the hydraulic head must be added (see Figure 9.5). This pressure is, of course, triangular beginning at the design height of the external water table and accumulating at the rate of 62.4 psf (3 kPa) per foot (0.3 m) of depth.

The effect of the inclusion of water pressure will affect the earth pressure as the buoyant weight of soil is now used for all soils below the water table. The unit weight of soil is described in Chapter 8.3 and 8.7 together with the effect of buoyancy. The effect of earth pressure, water pressure and a UDL surcharge can be seen in Figure 9.6.

There is disagreement in the design profession as to whether the active pressure on the wall extends into the toe of the wall. There can be no question that the pressure occurring from unbalanced water head (cases where dewatering inside the excavation does not affect the external water table) extends to the toe of the sheet pile or secant wall. However, some designers will also extend the active earth pressure down the outside of the wall for the width of the soldier pile to the base of the embedment (see Figure 9.7) while others do not. The stiffer the soils, the less the inclusion or exclusion of this pressure appears to affect the final design.

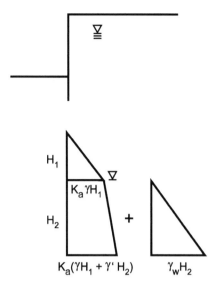

FIGURE 9.5 Active pressure with the effect of a water head.

Below the excavation, passive pressure develops on the excavation side of the wall because the wall is attempting to move into the excavation. Figure 9.8 shows the development of passive resistance. In cohesionless soils, the passive pressure can be calculated as a function of the depth of embedment. Chapter 8.5 outlines the development of passive pressure coefficient K_p. The actual calculation of passive pressure when applied to discrete wall elements, such as soldier pile toes, is discussed in Chapter 11.1.2.

The passive pressure (Pp) at any point in cohesionless soils is defined as

$$P_p = K_p \gamma d \quad (9.3)$$

Where d is defined as the depth of embedment
In c, Φ materials the passive pressure is defined as

$$P_p = K_p \gamma d + 2c \quad (9.4)$$

The passive pressure on the back side of the embedment (below the point of rotation) in a cantilevered pile toe is defined the same way. However, the depth used to calculate the pressure is the depth from the top of the wall instead of from the base of the excavation.

Some designers will ignore the effect of passive pressure in the first two feet below the excavation. This is because minor over-excavation or disturbance of material at the base of the wall may weaken the passive resistance of the soils.

274 EARTH RETENTION SYSTEMS

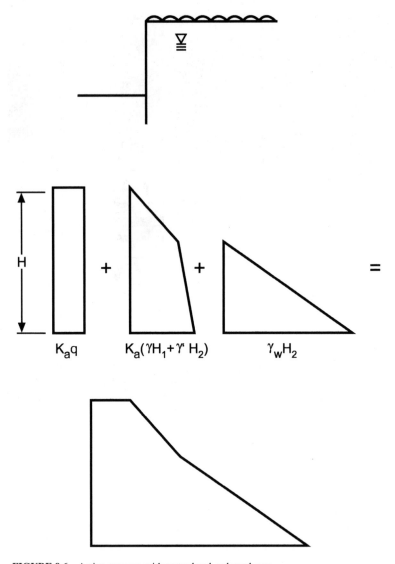

FIGURE 9.6 Active pressure with water head and surcharge.

In situations where the shoring is retaining slide materials, it is entirely possible that the passive pressure envelope necessary for wall stability may need to be depressed well below the base of the excavation. Certainly the passive pressure should not be assumed to be mobilized above any recognized slip plane.

FORCES ON WALLS

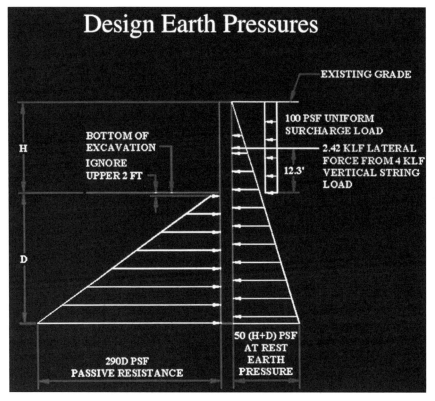

FIGURE 9.7 Active pressure extended to base of wall. *(Courtesy of Golder Associates, Inc. Redmond, WA)*

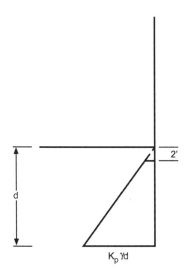

FIGURE 9.8 Passive pressure—cohesionless materials.

9.2 MULTIPLE LEVELS OF SUPPORT

Experience has shown us that earth pressure theory does not accurately predict loading experienced by multi strutted (or tied-back) shoring systems. Here, the general practice has been to use apparent earth pressure as the method to define loads when designing.

Apparent earth pressure derives principally from work done by Terzaghi and Peck and others on multi-strutted subway excavations. What came from this body of work was the development of a series of envelopes that predict the strut loads inherent in a shoring system. These envelopes are empirical and not rigorous in their derivation.

9.2.1 Sand

A rectangular pressure diagram is used (see Figure 9.9) where the pressure at any point on the wall above the base of the excavation is defined as

$$P = 0.65\ K_a\ (\gamma H + q) \tag{9.5}$$

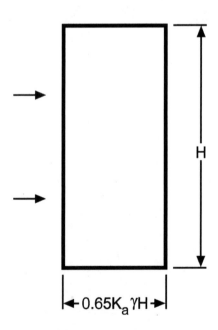

FIGURE 9.9 Apparent earth pressure–sand.

9.2.2 Soft to Medium Clays

A rectangular pressure diagram with a triangular top as shown in Figure 9.10 is used in cases where stability factor N is greater than 6.

$$N = \gamma H/c \quad (9.6)$$

In these cases

$$K_a = 1 - 4/N \quad (9.7)$$

If the excavation is underlain by a deep deposit of soft or sensitive clay,

$$K_a = 1 - 1.6/N \quad (9.8)$$

The pressure at any point below 0.25 H is defined as

$$P = K_a (\gamma H + q) \quad (9.9)$$

9.2.3 Stiff Clays

In stiff clays, or c, Φ materials, the diagram usually used is Figure 9.11.

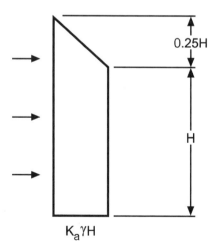

FIGURE 9.10 Apparent earth pressure–soft to medium clay.

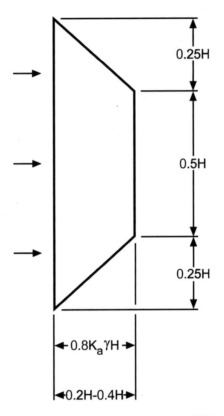

FIGURE 9.11 Apparent earth pressure–stiff clay.

The active pressure in the middle 50 percent of the diagram is defined as

$$P = 0.8 \, K_a \, (\gamma H + q) \tag{9.10}$$

where $0.2 < 0.8 K_a < 0.4$

Some authors depict the stiff clay diagram to show the point of maximum loading occurring 0.2H below the ground surface instead of 0.25H and continuing to within 0.2 H of the base of the excavation.

In cases where N is less than 4, $0.8 \, K_a$ will approximate the lower bound, while if N is between 4 and 6, $0.8 \, K_a$ will be 0.4.

In shored systems which have support from struts or tiebacks, the toe of the wall does not necessarily have a point of rotation, and as a result, no passive pressure moment couple, as outlined in the cantilever case, is developed. The horizontal earth pressures in the multi strut case are carried by the struts and the embedded portion of the wall. The embedded portion of the wall simply develops its lateral load carrying capacity with one passive pressure envelope. This passive resistance is best analyzed using Equations 9.3 or 9.4. This is only one of the many apparent inconsistencies in our design approach. Note that the lateral load on the wall (the active load) is being analyzed using apparent earth pressure while the passive pressure is from earth pressure theory.

It can be shown that for a soil mass to truly develop an active state it must undergo deflections in excess of 0.003H where 'H' is the depth of the cut. In contrast, while we would like to believe that at-rest pressure designs (see Chapter 8.6) can restrain movements to nil, in reality the types of movements for at-rest designs tend to be in the range of 0.001H. What Terzaghi and Peck seemed to have found was that when you had a braced excavation of considerable depth, the movements at the top were restrained by the strutting to something less than that which would permit fully active state development. As a result, the apparent earth pressure envelopes are dealing with a soil mass which is part way between at-rest and active and, not surprisingly we find, exhibit movements in the range of 0.001H-0.003H. Again, this makes sense. The total retained force in a rectangular apparent earth pressure sand diagram is about 1.3 times that of the total force from a triangular earth pressure theory diagram. The restraining forces are greater, so the deflection is less.

9.3 SINGLE STRUT OR TIEBACK

There is no agreement on what diagram to use for the single level of support. Some designers will use the triangular as indicated in the cantilever case (earth pressure theory). Others will use the sand diagram for multi levels. Still others will use the soft clay shaped diagram (truncated trapezoid) with K_a developed in accordance with Equation 8.1. The last two from apparent earth pressure.

Bulkheads and shoring where facilities are not sensitive to small movements, are usually designed with triangular pressure diagrams. Where large surcharges exist, or sensitive utilities or structures are involved, designers will usually use some form of apparent earth pressure diagram.

The triangular shaped diagram will increase the load on the embedded portion of the wall and will encourage a deeper placement of the strut or tieback. This will induce more lateral movement in the shoring system prior to anchor placement. The truncated trapezoid will increase the load on the tieback, and encourage the designer to place the tieback higher which reduces movement and results in less design load on the embedded portion of the wall. Most movement in a shored wall will occur during the excavation for, and installation of, the tiebacks. The higher the tieback, the less the excavation, and therefore the less the deflection. While, the use of high jacking forces in tieback stressing can sometimes recover deflections in a shored wall, it is much easier to prevent deformations by not allowing them to occur in the first place.

9.4 SOIL NAILING

Although the previous methods of analysis have emphasized the wall as a load bearing structure with external loads impressed on the wall, soil nail walls, like Mechanically Stabilized Earth Walls (MSE) are really examples of artificially strengthening the soils immediately behind the wall, such that the soil mass retains itself. Using limit equilibrium analysis, the internal forces (normal and shear) are considered within the altered (reinforced) soil mass affected by the adjacent excavation.

In limit equilibrium analysis, the soil mass analyzed is defined by the ground surface, the wall face and a failure plane. The mass is cut into small slices and forces are reconciled within each slice and across the failure plane (see Figure 9.12). This is why Limit Equilibrium is often called the "Method of Slices." It has been accepted that the failure plane will be a curved surface, often approximated as a circular, parabolic or log spiral surface (see Figure 9.13). Various failure planes are tried to determine the critical failure plane. Designs performed using Limit Equilibrium Analysis tend to exhibit movements in the same range as apparent earth pressure (0.001H-0.003H).

Limit equilibrium analysis is sometime used as a check for global stability in tied-back excavations where constraints on the length of tiebacks force designers to utilize unconventional no-load zones.

9.5 BASE STABILITY

When designing a shored excavation, the designer must ensure that the excavation will be stable when completed and not suffer from base heave. Base heave is the phenomenon which occurs when the overburden pressure of the soil and surcharge outside the wall overcomes the bearing capacity of the soils within the excavation. The soils then fail and flow under the wall and up into the excavation (see Figure 9.14).

Base heave must be considered in soft or medium clays, or in water bearing loose sands and silts where dewatering inside the excavation will cause unbalanced hydrostatic pressures. In soft or medium clays the factor of safety (FS) against basal heave can be calculated as follows:

Step 1 Determine bearing capacity factor Nc from chart (Figure 9.15)

Step 2

$$F.S = N_c \cdot c/(\gamma H + q) \qquad (9.11)$$

where 'c' is the shear strength of soil below the base of the cut.

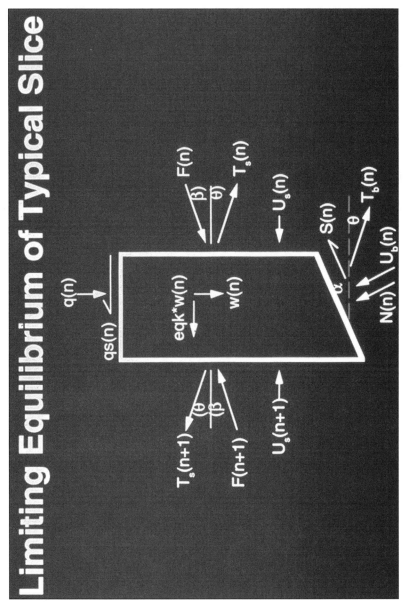

FIGURE 9.12 Method of Slices Analysis. *(Courtesy of Golder Associates, Inc. Redmond, WA)*

282 EARTH RETENTION SYSTEMS

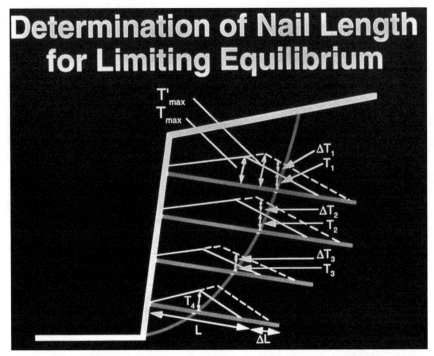

FIGURE 9.13 Slip planes with soil nails. *(Courtesy of Golder Associates, Inc. Redmond, WA)*

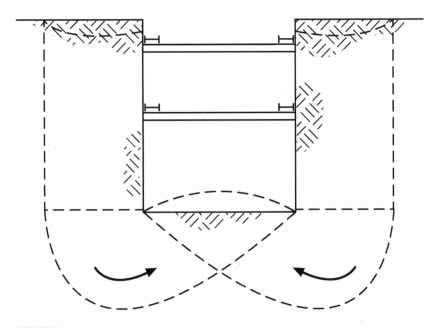

FIGURE 9.14 Basal heave.

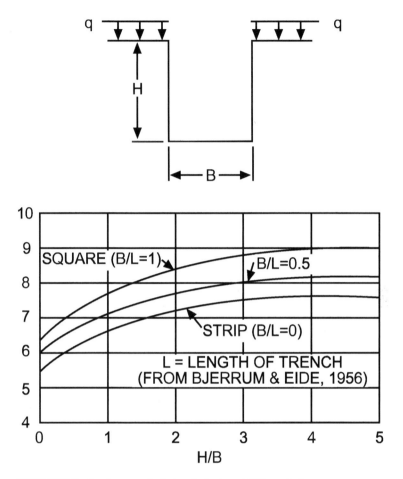

FIGURE 9.15 Bearing capacity factors for bottom stability analysis.

If the factor of safety is less than 1.5, in order to safely complete the excavation, it may be necessary to extend the shoring wall below the excavation in order to lengthen the failure path. Many designers believe the factor of safety cannot be increased by the use of a flexible retaining system such as sheet piling, and that it may require a more rigid wall such as a slurry wall or a secant wall in order to improve the performance of the excavation.

If dewatering of a shored excavation in cohesionless soils is carried out and the shoring does not extend to a layer of soil which will cut off the flow of water from outside the excavation, the designer must ensure that basal instability will not occur because of flow of water under the wall and upward toward the base of the excavation. This eventuality can be analyzed by developing a flow net to analyze the stability of the base of the excavation.

9.6 SURCHARGE

Surcharge loading can occur in a number of ways. Construction machinery or materials may be staged adjacent to the excavation (see Figure 9.16). Adjacent structures will impose surcharges through their footing loads. Traffic adjacent to an excavation will impose surcharge loads. If shoring is designed such that the excavation slopes upward above the top of the shoring, either naturally or as a result of cutting operations, the slope certainly will surcharge the wall and must be considered.

9.6.1 Construction Material and Equipment

Designers often will use a surcharge of 200 psf (10 kPa) as a uniformly distributed load (UDL) for design purposes, or sometimes 2 feet (610 mm) of soil (about 240 psf (11.5 kPa)). This would appear satisfactory in most cases. If we assume that a vertical load will distribute itself by spreading at a rate of 1:1 (for every foot below the point of loading the zone of influence increases by one foot in each direction), a line of ready mix concrete trucks fully loaded, parked end to end 4 feet (1.2 m) from the edge of the excavation will exert of vertical surcharge load of 160 psf (7.7 kPa) at a depth of 4 feet (1.2 m) from the ground surface. At this point the surcharge load contacts the wall. As you can see, 200 psf (10 kPa) will cover most cases.

The designer should, however, ensure that the contractor is aware of the 200 psf (10 kPa) limit in the design. If heavy cranes are adjacent to the excavation or extreme stockpiling of materials are anticipated, then this figure should be adjusted. Similarly, if wheel loads are anticipated closer than 4 feet (1.2 m) from the shored face, the surcharge should probably be analyzed as a point load rather than a UDL.

9.6.2 Traffic Loading

Since traffic loading would almost never be greater than fully loaded ready mix trucks end to end (see Chapter 9.6.1) it can be assumed that 200 psf (10 kPa) would be satisfactory for most traffic loading. The exception is rail traffic which should be analyzed using railway loadings which have been codified in methods such as Cooper E-80.

9.6.3 Adjacent Structure Loading

Adjacent structures should be analyzed to determine their footing loading. The amassing of this information is discussed in detail in Chapter 7. The effect of the footing loading can then be determined by a Boussinesq analysis to determine the vertical and horizontal component of the footing loading on the adjacent shoring.

FIGURE 9.16 Surcharge loading from placement of an erection crane. (*Courtesy of Condon-Johnson & Associates, Inc. Seattle, WA*)

9.6.4 Backslopes (Cut or Natural)

Two methods exist to analyze the effect of backslopes on shored excavations. Coulomb analysis has been developed which will alter K_a for the effect of the backslope. It can be shown that, when reduced for vertical wall inclination and zero backfill

$$K_a = \frac{\sin^2(90+\Phi)}{r} \tag{9.12}$$

$$r = \left(1 + \sqrt{\frac{\sin\phi\sin(\phi-\beta)}{\sin(90+\beta)}}\right)^2 \tag{9.13}$$

where
$\quad\Phi$ = angle of internal friction
and
$\quad\beta$ is the angle of the slope behind the wall (see Figure 9.17)

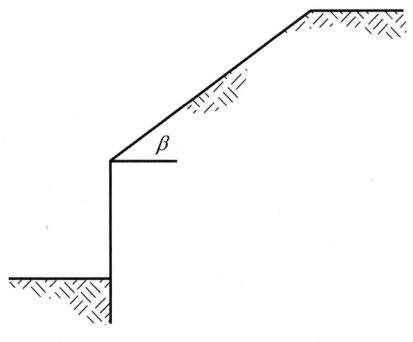

FIGURE 9.17 Backslope behind shoring.

When viewing this messy formula, it is not surprising that designers have found other ways of dealing with backslopes above retaining walls.

As an alternative, designers will model the slope behind a shored wall as exerting a surcharge on the wall equal to a function of height of the backslope. In Figure 9.18 the designer placed a surcharge equal to 50 percent of the height of the backslope to approximate the effect of the backslope on the shoring. The figure 50 percent should not be taken as gospel and the designer should take into account the slope angle to determine the amount of surcharge to assume. One way of doing this is to calculate the total weight of soil surcharge which would fall within a line drawn at an angle of $45° - \Phi/2$ upward from the base of the excavation (the active zone or Rankine wedge) (see Figure 9.19). This weight is then distributed over the distance behind the wall defined by the edge of the wall and the point of intersection of the Rankine wedge with the level of the top of the wall as a UDL at the top of the wall.

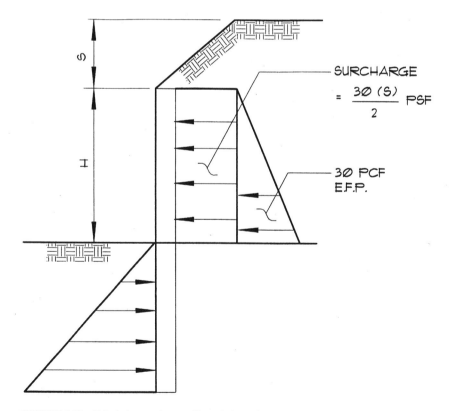

FIGURE 9.18 Calculating surcharge effect of slope above shoring wall. *(Courtesy of CT Engineering, Seattle, WA)*

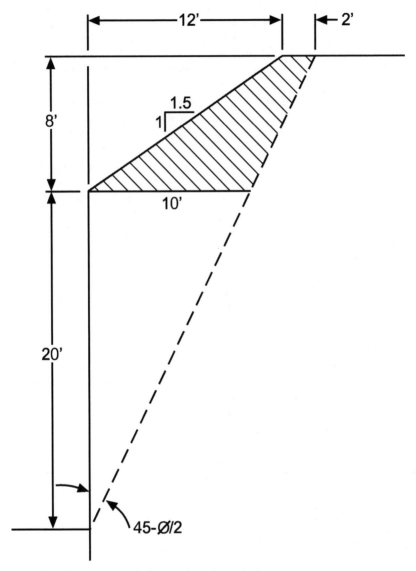

FIGURE 9.19 Surcharge calculation—alternative method.

EXAMPLE 9.2 CALCULATE EQUIVALENT SURCHARGE AS A RESULT OF BACKSLOPE.

Backslope height = 8 feet
Backslope angle = 1.5/1
Wall height = 20 feet
Unit weight of soil = 120 pcf
Φ = 30 degrees

The active zone, defined by a line drawn upward from the base of the excavation at 45° − $\Phi/2$ contacts the ground surface 14 feet back of wall.

Weight of soil within this wedge (see Figure 9.19):

$$8 \bullet (2 + 10)/2 \bullet 120 = 5760 \text{ lb.}$$

Surcharge to be applied over 10 feet (intersection point of Rankine Wedge with top of wall plane behind wall)

Surcharge is 5760/10 = 576 psf

CHAPTER 10
FAILURE MODES OF SHORING

In order to understand the design of shoring systems, it is necessary to have a clear understanding of the types of failures which can occur to a shoring wall. These failures include

- Structural failure of some component of the shoring
- Geotechnical failure of some soil component in contact with the shoring
- Facial instability
- Basal instability
- Global instability

10.1 STRUCTURAL FAILURE

Structural failure of a shoring system occurs when some portion of the built system is not sufficiently strong to withstand the imposed loads. The overload could be the result of an inaccurate estimate of the imposed load, or may be caused by a geotechnical failure which then overloads a structural portion of the wall.

A failure in cantilever (see Figure 10.1) will occur when the cantilever portion of the structure (above the tieback, strut or raker level), is not sufficient to withstand the imposed loads and fails either in bending or in shear. This type of failure could also occur prior to the installation of the first row of tiebacks if the cantilever capacity of the piling is exceeded.

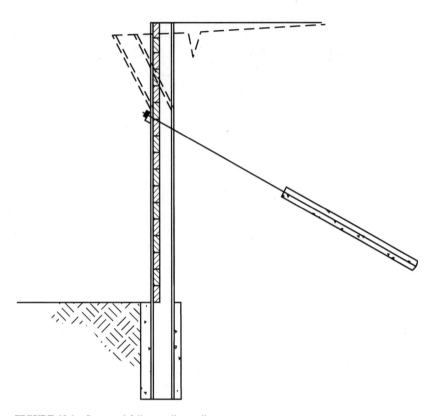

FIGURE 10.1 Structural failure—pile cantilever.

Failure in midspan of the wall piling between tieback or strut levels will occur if bending capacity is exceeded (see Figure 10.2).

A structural failure could also occur if the connection between the wall and the tieback fails. A similar failure would occur if a strut or raker failed in buckling, or if a waler failed in bending. A failure of this type will overload the piling in bending or overload other tiebacks, struts or wall embedments. In the case of discrete piling elements such as soldier piles or secant piles. The failure of one tieback could can result in a zipper type of failure where loads are thrown onto tiebacks above or below the failed tieback, causing overload and subsequent failure. In walered systems the overload may be transferred laterally to adjacent tiebacks causing failure either in the waler or the adjacent tiebacks.

Failure of a tieback tendon from excessive tension will also result in overloading adjacent tiebacks or cause the bending resistance of the wall piling to be exceeded. The resultant failure would be similar to a failure of a tieback connection (see Figure 10.3).

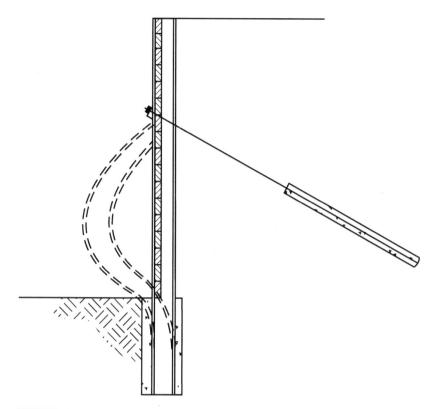

FIGURE 10.2 Structural failure—pile midspan.

Structural failures usually result in catastrophic wall displacements, which often bring about unacceptable movements in the adjacent soil mass and damage to any facilities which are located within the affected soil mass.

10.2 GEOTECHNICAL FAILURES

Geotechnical failures occur when the soil strength is not sufficient to resist the imposed loads applied by the constructed portions of the wall. Geotechnical failures usually end up redistributing imposed loads to other portions of the wall often with catastrophic results. Some types of geotechnical failure are discussed below.

Walls or wall elements may fail by sinking, when the downward component of the tieback load or other imposed vertical loads is greater than the pile bearing capacity in friction and end bearing. This sinking will allow the pile to rotate forward as the tieback becomes detensioned, causing movement of the retained soil mass and subsequent damage to adjacent structures (see Figure 10.4).

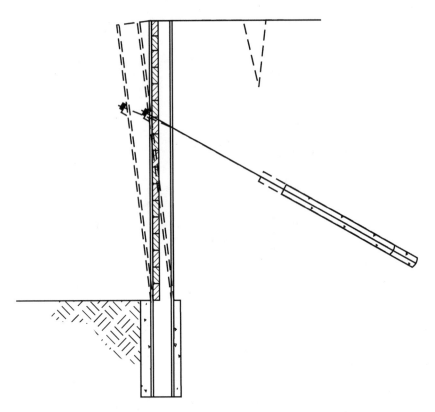

FIGURE 10.3 Tieback failure—bond failure. Note if the tendon or tendon/pile connection fails the movement will be similar although the bond zone will not displace as in a geotechnical failure. *(Courtesy of Golder Associates, Inc. Redmond, WA)*

Walls or wall elements may move vertically upward if the friction on the wall is insufficient to withstand the uplift caused by inclined struts or rakers. This upward movement will allow rotation of the pile which permits movement of the soil mass. Unlike the downward movement of piling under tieback loading which can be self limiting, once uplift movement begins, it will accelerate unless stopped immediately (see Figure 10.5).

Failure of the tieback bond, permitting slippage of the tieback, will result in overloading of adjacent tiebacks and/or bending failure in the wall elements. The result will be similar to that discussed when tieback connections fail.

If soil nails have inadequate bond, an under-reinforced soil mass will ensue which will manifest itself in a wracking (Figure 4.5) of the soil mass and a progressive type of failure.

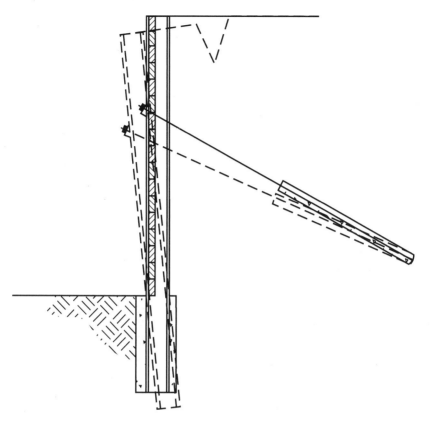

FIGURE 10.4 Settlement of pile toe. Note that this permits rotation of the pile forward.

Bearing capacity failure of raker footings will result in the inward rotation of the wall piling. This again will permit excessive movement of the soil mass and subsequent damage to adjacent facilities (see Figure 10.6). In addition, if walers are involved in the system, load will be shed through the waler to adjacent rakers with possibly catastrophic effects

A passive resistance failure at the base of the shoring wall will permit rotation of the wall inward at the toe. This movement can permit excessive soil mass movement causing damage or may result in overloading of the wall piling in bending and subsequent structural failure (see Figure 10.7).

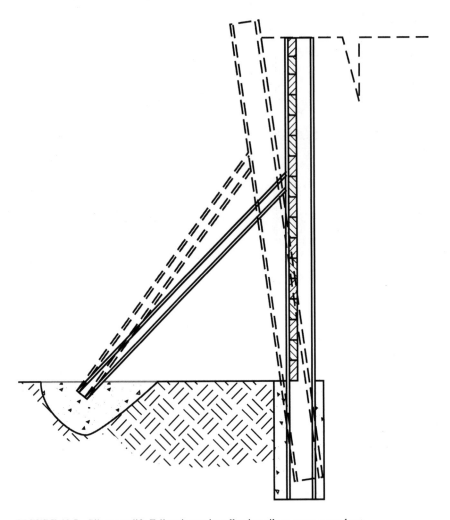

FIGURE 10.5 Pile toe uplift. Failure in tension allowing pile to rotate up and out.

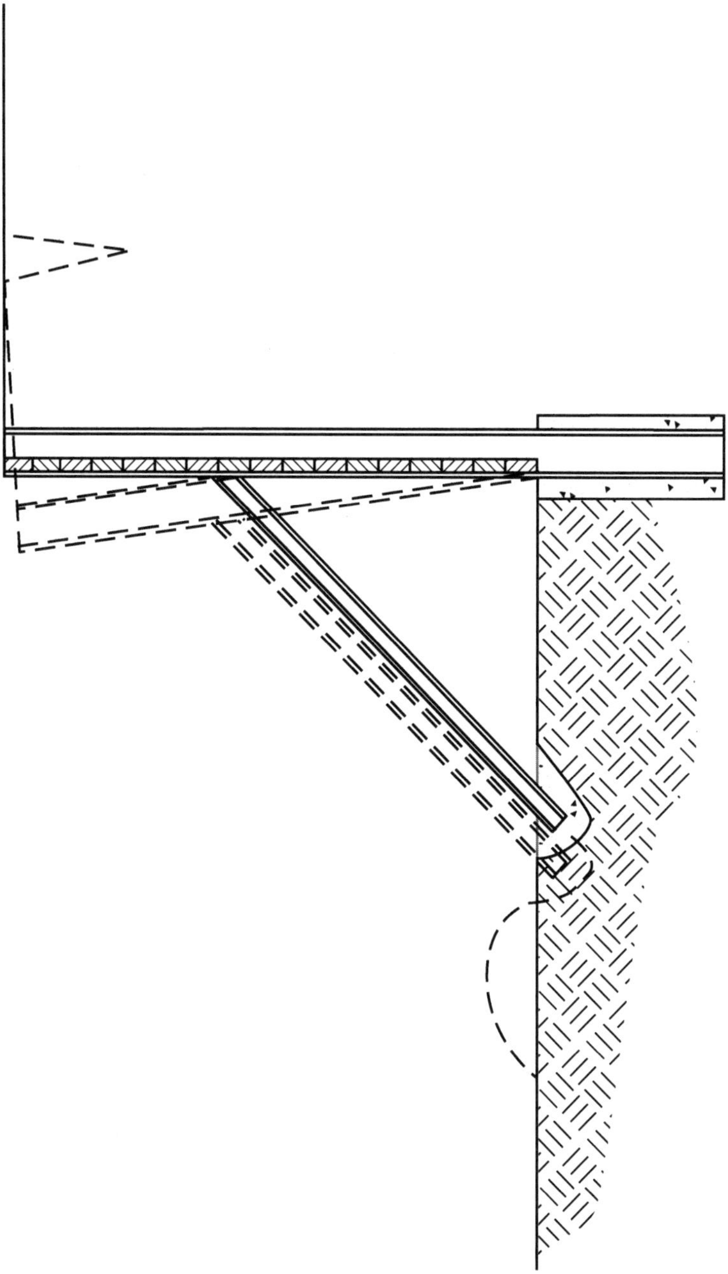

FIGURE 10.6 Raker bearing capacity failure.

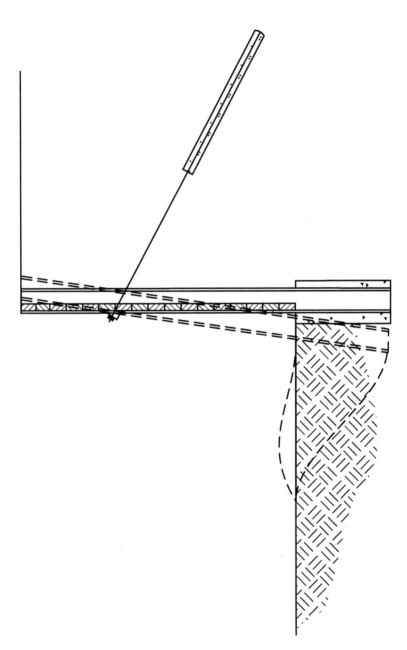

FIGURE 10.7 Passive resistance failure of pile (soldier pile or sheet pile).

10.3 FACIAL INSTABILITY

Facial instability occurs when soils in direct contact with lagging systems, or soil nailed systems, do not exhibit sufficient stand up time to prevent sloughing. If sloughing occurs at the exposed face of a cut being prepared for lagging or shotcrete, it can affect the lagging or shotcrete already installed above the base of the excavation. Once sloughing occurs below the wall fascia (lagging or shotcrete), it has a tendency to progress upwards until it reaches the ground surface in a process known as "chimneying."

Facial instability causes a loss of frictional contact between the soil mass and the wall. In soldier pile and lagging systems, this loss of friction will place added load on structural elements of the wall and could cause geotechnical failure of the pile toes. Loss of friction behind a lagging system will direct all vertical loads to the soldier pile toes. Loss of friction behind a soil nail fascia will force the nails to carry the wall fascia weight as a cantilever load. This may result in a structural failure of the nail and a subsequent dropping of the wall fascia.

Both results will cause significant movement of the soil mass behind the wall with subsequent damage to adjacent facilities. Even if geotechnical or structural failures do not occur, the loss of soil can eventually create settlements which risk damage to adjacent sensitive installations.

10.4 BASAL INSTABILITY

Basal instability is the tendency of the base of the excavation to heave or boil when excavated. Boiling occurs when the water level is higher outside the excavation than inside the excavation and a flow of water is possible. This water flow will disturb the soil and cause a loss of contact between soil particles. In extreme cases, it is evidenced by a bubbling of the base of the excavation (hence the term boiling). This boiling of the basal soils of the excavation disrupts the bearing capacity and passive resistance of the soil and tends to worsen with time as the water flow creates piping channels to permit its flow in ever increasing amounts (see Figure 10.8).

Basal instability does not necessarily require water flow to occur. If the difference in overburden pressures between the inside and outside of the excavation is sufficient to overcome the shear strength of the affected soils, the soils will flow from the outside of the excavation under the wall and up inside the excavation. This phenomenon will destabilize the wall, disrupt the soil mass outside the excavation with accompanying damage to adjacent facilities and diminish the bearing capacity of the basal soils within the excavation (see Figure 10.9).

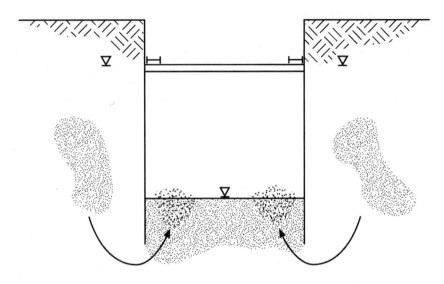

FIGURE 10.8 Boiling.

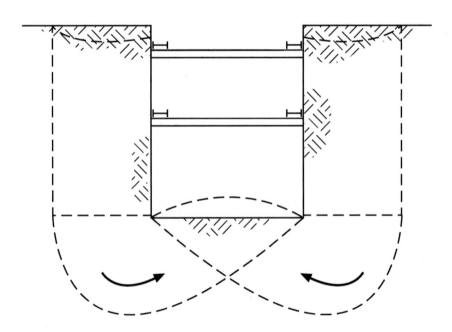

FIGURE 10.9 Basal heave.

10.5 GLOBAL INSTABILITY

In cases of variations in strength of materials, or the incidence of slide planes, or abnormal geometry, the embedment of tiebacks behind the active zone conventionally known as the Rankine Wedge (see Chapter 8.4) may not be sufficient to ensure that a global failure does not occur. Figure 10.10 is indicative of the type of failure which may occur if the entire soil mass is subject to movement. This failure is catastrophic and results in a great deal of damage to adjacent facilities as well as the wall itself. A global failure is commonly depicted as a circular type of failure but also may occur if a shallow slip plane is activated causing a large slab to move.

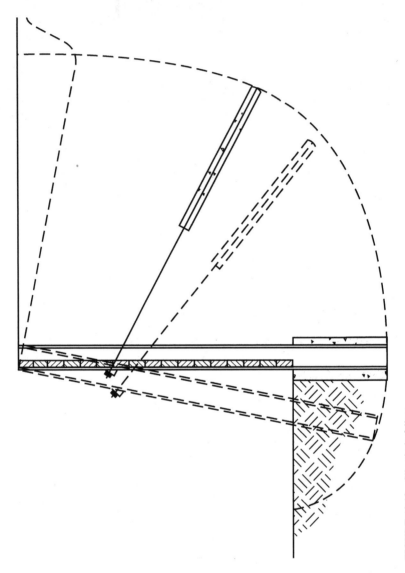

FIGURE 10.10 Slip circle failure.

CHAPTER 11
DESIGN METHODS

The actual design of a shoring wall is usually done with the use of computer software developed to specifically perform the calculations. These calculations are particularly suited to solution by computer as they tend to be iterative in nature. It is, however, important to understand the basis of the calculations so that checks can be made to assure that computer solutions are credible.

This chapter will not provide actual design examples. Many examples are detailed in Chapter 17 as well as the reference texts outlined in the Bibliography. This chapter will, however, detail design methods which might be followed to design the various components of a shoring system.

11.1 CANTILEVER

11.1.1 Cantilever—Continuous Wall

By continuous wall, we mean a wall in which the section is continuous throughout, such as a sheet pile wall, slurry wall, or secant wall with intermediate piling at the same depth as the primary piles (see Figure 3.38). The design of such a wall might be performed in the following manner:

Step 1. Define parameters of wall and soil

- Height of wall: H
- Soil parameters: c, Φ
- Surcharge: UDL, Boussinesq
- Unit weight of soil: γ
- Water table on either side of wall

Step 2. Develop K_a, K_p, q

Step 3. Select a pressure diagram (see Figure 11.1). Develop the pressure diagram based on a unit length of wall (1 foot length).

- See active pressure formulas Equations 9.1, 9.5, 9.9, and 9.10 for P.
- See passive pressure formulas Equations 9.3, 9.4 for R_2. In cases where a hydrostatic head exists on either side of the wall (even if it is not equal), deduct one from the other so that only the difference is used.
- Assume that the lower passive pressure, outlined as R_3 extends to a depth of three feet below the bottom of R_2 (Point A). Center R_3 at 2 feet below bottom of R_2 (Point A).

Step 4. Calculate the capacity of the embedment for a given depth 'd' assuming a Factor of Safety (FS) of 1.5

$$R_u = K_p \, \gamma d^2/2 + 2cd \qquad (11.1)$$

(use γ' if below the water table)
Where R_u is the ultimate capacity and R_2 is the design capacity after application of FS.

$$R_2 = (K_p \, \gamma d^2/2 + 2cd)/1.5 \qquad (11.2)$$

Note: Geotechnical reports will often give a value of passive resistance in terms of equivalent fluid pressures. In other words, the figure stated is equal to $K_p \gamma H$ and is quoted as xH pcf. The designer must check as geotechnical engineers will often include a factor of safety of 1.5 in this figure. It is important to clarify this matter as it will affect not only the depth calculation but also the section modulus (see Step 6). Figure 11.2 outlines typical earth pressure recommendations from a geotechnical report.

Step 5. Balance moments about R_3. This will involve taking various depths of 'd' and balancing moments from P_t and R_2. Establish a depth 'd' of R_2. Embedment will then be 'd' plus 3 feet for R_3.

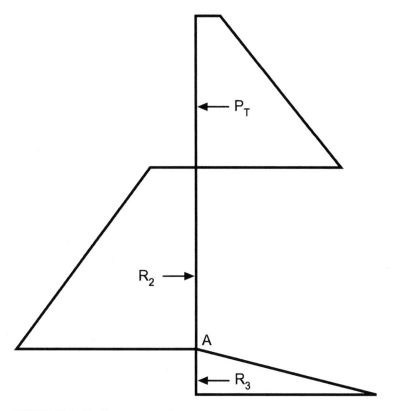

FIGURE 11.1 Cantilever pressure diagram.

Step 6. Check R_3. Set the Passive Pressure on back side of pile to 2c at the base of R_2 (Point A) Maximum Passive Pressure at Point A plus 3 feet is

$$P_p = (K_p \gamma d) + 2c \tag{11.3}$$

In this case 'd' is equal to the depth from top of wall.

$$R_3 = 3((K_p \gamma d^2/2) + 2c) \tag{11.4}$$

And assuming a F S of 1.5

$$R_3 \text{ allowable} = K_p \gamma d^2 + 4c \tag{11.5}$$

Step 7. Find the point of zero shear. This will equate to the point of maximum moment.

306 EARTH RETENTION SYSTEMS

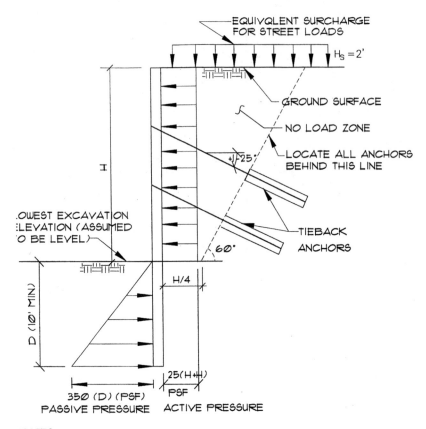

NOTES:
1. THE REPORT SHOULD BE REFERENCED FOR SPECIFICS REGARDING DESIGN AND INSTALLATION
2. ACTIVE PRESSURES ACT OVER THE PILE SPACING
3. PASSIVE PRESSURE ACT OVER TWICE THE GROUTED SOLDIER PILE DIAMETER OR THE PILE SPACING, WHICHEVER IS SMALLER.
4. IT IS ASSUMED THAT NO HYDROSTATIC PRESSURES ACT ON THE BACK OF THE SHORING WALLS.
5. CUT SLOPES, STRUCTURES, OR OTHER SHORING WALLS POSITIONED ABOVE OR BEHIND SHORING WILL EXERT ADDITIONAL PRESSURES ON THE SHORIG WALL.

FIGURE 11.2 Typical Geotechnical recommendations for shoring design. *(Courtesy of CT Engineering, Seattle, WA)*

Note: Use R_u without factor of safety. All steel sections will, by code, have a factor of safety of approximately 1.5 to 2 depending on the code used. If R_2 is used, a duplication of factors of safety will occur.

Step 8. Calculate the maximum moment and select the appropriate steel section. Check basal stability (see Chapter 9.5). If the factor of safety is less than 1.5, serious consideration should be given to excluding the use of sheet piling for this application unless soil improvements are anticipated, such as jet grouting or deep mixing which will improve the basal stability of the excavation. Deepen the embedment of the wall as necessary to ensure basal stability.

11.1.2 Cantilever—Discontinuous Wall

Discontinuous walls include soldier pile and lagging walls, tangent pile walls, and secant pile walls where the embedment is not constant as indicated in Figure 3.39.

Steps 1 and 2. Same as continuous wall.

Step 3. Develop the active pressure diagram based on loads on one bay (the distance from one pile to the next) of shoring. Example, if soldier piles are at 'b' foot centers the active pressure at h will be as follows in cohesionless soils

$$P\ @h = b \cdot K_a(\gamma h + q) \tag{11.6}$$

where b is equal to the bay spacing and h is the depth at the point of calculation.

Develop the passive pressure diagram based on loads on one single soldier pile. It was demonstrated by Broms in 1964 in a series of papers about laterally loaded piles that piles would develop passive pressure on a width of up to three times their width. When this principle is applied to soldier piles, if the soldier pile is placed in a 2 foot (610 mm) diameter drilled hole, the effective passive pressure would be over a width of up to 6 feet (1.8 m). Some designers will use 2x or 2.5x. The width used should never be greater than the bay width (b) of the soldier piles.

$$P_p @\ d = 3 \cdot B \cdot ((K_p \gamma d) + 2c) \tag{11.7}$$

where 3 is the multiplier suggested by Broms.

B is the diameter of the drilled hole in which the pile is placed. In the case of driven soldier piles, use the width of flange of soldier pile.

Note: If the designer concludes that the active pressure must be carried into the toe of the soldier pile (see Chapter 9.1), then the active pressure is applied to the back of the soldier pile in the embedment zone on a width of 1x the diameter of the soldier pile embedment only.

Steps 4 through 8. The same as continuous wall analysis. If a basal stability check indicates that the factor of safety is less than 1.5, the use of discontinuous walls should not be considered for this application unless the base of the excavation is to be improved by some method such as jet grouting or deep mixing to improve its resistance to basal heave.

11.2 MULTIPLE TIEBACKS (OR STRUTS)

In the case of multiple levels of support, the question is always, "How do we optimize the system?" Is the optimum system the one which has the lightest piles, or does it have the fewest struts or tiebacks? The answer is, "it depends." It depends upon the cost of strutting or tiebacks, and it depends upon the cost of soldier piles. It depends upon whether the local authority insists on destressing or removal of tieback tendons and does this destressing affect the forming system for the concrete work inside the shoring? However, the most economical system usually involves the lightest vertical members (soldier piles or sheet piles) with tiebacks in the range of 100 Kips to 200 Kips (45-90 T). To accomplish this, a system in which the bending moments in all aspects of the design are balanced is necessary.

11.2.1 Design of Multiple Level of Support

Step 1. Define the wall and soil parameters per Chapter 11.1.

Step 2. Determine values K_a, K_p and q.

Step 3. Develop the active pressure diagram (see Figure 11.3). Select the appropriate diagram from Chapter 9.2. Assume a depth of toe and corresponding passive pressure diagram. Use a continuous embedment or discontinuous embedment model as appropriate (see Chapters 11.1.1 and 11.1.2).

Step 4. Calculate the load on each level of support by splitting the distance between supports. Check moment balance by taking moments about the upper strut (R_1). Rebalance support loads to achieve moment equilibrium.

Step 5. Find the horizontal load on the pile toe and check the initial assumptions. Change toe depth assumption if necessary.

Step 6. Calculate moments at each support. Plot the moment at each support on moment diagrams such as Figure 11.4.

Step 7. Calculate the simply supported moment (m = $wl^2/8$) between each support. Where 'w' is the active pressure from Equations 9.1, 9.5, 9.9, or 9.10.

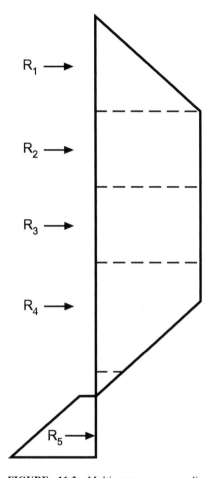

FIGURE 11.3 Multi-strut pressure diagram—first guess on tieback loads is based on splitting the distance equally between struts to calculate the contributory load. Note R5 is the pile toe.

Add this moment to moment diagram (see Figure 11.4). This should create a system of positive moments at each support with a negative moment between the supports. An optimized system will have positive moments of the same size as the negative moments.

Note: In order to balance the moments, it may be necessary to increase or decrease the spacing between supports.

EARTH RETENTION SYSTEMS

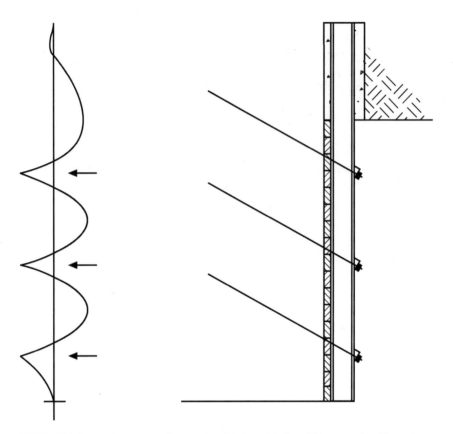

FIGURE 11.4 Bending moment diagram. A well balanced design will have equal positive and negative moments.

Step 8. Calculate the bending moments prior to each support installation. Assume two feet (610 mm) of overexcavation in each case.

Note: Depth H for solving the equations in Step 7 when checking these intermediate steps is only the depth of excavation at the time of checking, not the entire depth.

Step 9. Rebalance again to achieve similar moments in each cantilever, midspan, over strut, and pre-strut case.

Note: Some designers will use a lower F S for intermediate step checks as the duration of each step is quite short.

Step 10. Check the vertical component of the tieback load together with any imposed vertical loads. Distribute in accordance with the method selected from Chapter 11.6. Check the toe depth to ensure sufficient capacity.

Step 11. Multiply any moment derived from earth pressure by 0.8.

Note: This is called a moment reduction factor which is referenced in Peck, Hanson & Thorburn, FHWA Manual Vol II, and Terzaghi & Peck and represents the apparent earth pressure attempt to recognize the arching of the soils induced by the flexure of the system between support locations. This moment reduction should not be extended to moments resulting from hydrostatic loads (e.g., water behind sheet piling), because water does not arch.

Select the vertical member (soldier pile, slurry wall section, sheet pile) based on the reduced bending moment together with vertical imposed load. Use beam column analysis if vertical loads are significant.

11.2.2 Alternative Toe Design Method

Peck, Hanson & Thorburn noted that "a point of contraflexure" occurs very close to the base of the excavation in the vertical member. Some designers have interpreted this to mean that a point of zero moment occurs at or near the base of the excavation. To model this, they will place a hinge at the base of the excavation (see Figure 11.5). This hinge has two effects. It will change the moment distribution slightly in the vertical member and it makes the analysis much easier as it removes one degree of indeterminacy.

Because of the hinge, the location of the toe resistance is known and so it is possible to calculate strut loads by balancing moments without having to constantly recalculate the effect of the toe. Once the strut loads are known (and therefore the toe resistance), the toe depth can be calculated with one calculation rather than a series of iterations.

11.3 SINGLE STRUT OR TIEBACK

The analysis of a single strut is a simple operation. What is more difficult is deciding what to optimize. Depending on which diagram you choose for the active pressure, a design approach which optimizes bending moment in the vertical element may encourage far more movement than can be tolerated by the adjacent facilities (see discussion in Chapter 9.3). The following approach is for moment optimization and therefore economy of steel, but the designer must balance this against acceptable movements when making selections.

Step 1. Define wall parameters per Chapter 11.1.

Step 2. Determine K_a, K_p, q.

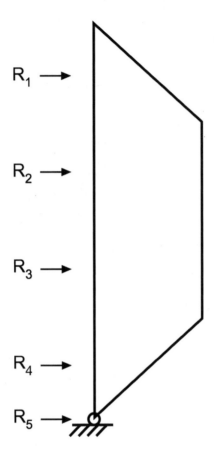

FIGURE 11.5 Some designers use a hinge at base of excavation for analysis. This eases the calculation of toe capacity and bending moment.

Step 3. Develop active pressure diagram (see Figure 11.6). At this point the designer must choose from the options presented in Chapter 9.3. Assume depth of toe and corresponding passive pressure diagram. Choose a continuous or discontinuous embedment model (Chapters 11.1.1 and 11.1.2) based on the type of wall being considered.

Step 4. Sum moments about the strut or tieback (R_1) and check the adequacy of the toe assumptions. Recalculate if necessary.

Step 5. Calculate the strut load by deducting R_2 from P_t (total Load).

DESIGN METHODS 313

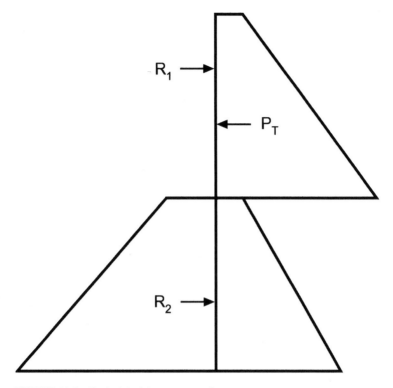

FIGURE 11.6 Single tieback/strut pressure diagram.

Step 6. Calculate the maximum bending moment by finding the point of zero shear. Ensure that the maximum moment in this case is below the strut level.

Step 7. Calculate maximum moment in the vertical element prior to installation of strut (assume two feet of excavation below strut). Use the method indicated in Chapter 11.1.

Step 8. Adjust the strut location to bring these two calculations into equilibrium.

Note: Lifting the strut will decrease the moment in the pre-strut case and increase the moment in the strutted case. Lowering the strut will have the opposite effect.

Step 9. Modify the maximum moment for the selection of the vertical member by 0.8. (see discussion in Chapter 11.2.1, Step 11). This analysis method can also be simplified by the method of assuming a hinge at the base of the excavation as discussed in Chapter 11.2.2.

11.4 DEADMAN ANCHORAGE

To design a deadman, the designer must be concerned about the depth of burial of the deadman and its distance behind the wall. In order to maximize the capacity of a deadman, the burial must be at a sufficient distance behind the wall to allow the deadman to develop its full passive resistance without conflicting with the active zone behind the wall (see Figure 11.7). Note that the active zone, or no load zone as it is referred to when designing tiebacks, follows the Rankine line of $45° - \Phi/2$ sloping upward from the base of the excavation, but is laterally shifted by some function of H. In Figure 11.7 this line intersects the ground surface at point A.

Some geotechnical designers will use H/3 while others favor H/4 or H/5. Probably the most commonly used is H/5. While some practitioners feel that this is an attempt to ensure that no load is dispersed within the "no-load zone," with the offset provding an added safety factor, it should be noted that the limits defined by this envelope are similar to the curved failure plane predicted by limit equilibrium analysis (see Chapter 9.4).

When a load is applied to a deadman, the anchorage resists through the development of a passive pressure wedge. As discussed in Chapter 8.5, this wedge can be defined by a line from the base of the anchorage sloping upwards at an angle of $45° + \Phi/2$, and in Figure 11.7, intersects the ground surface at point B. It is important that point B always be placed outside the active zone. In other words, Point B should always be behind Point A.

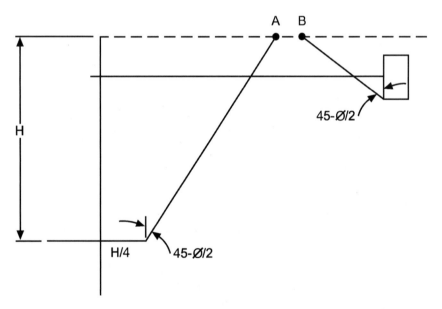

FIGURE 11.7 Deadman arrangement. Note that the passive pressure wedge of the anchorage is outside the active zone behind the wall.

The capacity of the anchorage is defined in terms of passive pressure principles, and as an example, the unit capacity of the anchorage in a cohesionless material is

$$P = K_p \gamma d^2/2 \qquad (11.8)$$

where 'd' is the depth of burial of the anchorage.

If the anchorage is to be continuous, then the anchorage must be designed to span in bending between the points of connection with the anchors. If the anchorage is to be a discrete block for each anchor, then the capacity of the anchorage may be influenced by some of Brom's thinking in that the width of the anchorage used for design should be increased in a manner similar to that used for Soldier piles in 11.1.2, Step 3.

No specific factor is recommended here, as each case will be affected by the depth of burial and the width of the anchorage proposed. The anchorage must be designed in cantilever bending about its point of attachment.

11.5 TIEBACKS

Tiebacks, whether they are anchored in soil (soil anchors) or rock (rock anchors) are designed so that they develop their capacity in friction along some portion of their length. While past practice at times relied on belled ends of drilled anchors and/or anchor plates to develop a passive cone of resistance, current thinking holds that most anchors will develop a frictional load along a defined length.

The capacity of an anchor must be developed behind the "no-load zone" discussed in Chapter 11.4 in order to assure that a global type of failure (pile and anchor move together) does not occur (see Figure 11.8). Given the load derived from the analysis of the vertical elements (Chapters 11.2 and 11.3) the tieback can be designed.

While the no-load zone discussed heretofore has been defined in terms of ϕ and H, it should be noted that in cases of unstable hillsides or ancient slides, the no-load zone may need to be defined by geotechnical evaluation to ensure that anchorage does not occur in unstable materials. These situations will override the simplified no-load zones exhibited here. In fact, the author has participated in projects where the no-load zone for a 30 foot (9.1 m) high wall was as long as 180 feet (55 m).

Step 1. Determine the horizontal load on the tieback.

Step 2. Determine the angle of declination desired. At times this is influenced by the depth of utilities or other underground facilities which may be in the vicinity. Current thinking usually requires that the tieback pass over a utility by five feet (1.5 m) or under by three feet (0.9 m) in order to assure missing it. Tieback angles are also influenced by the desire of the designer to anchor in a specific strata and

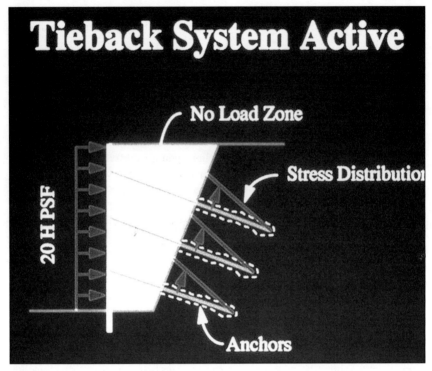

FIGURE 11.8 Active zone (no load zone). *(Courtesy of Golder Associates, Inc., Redmond, WA)*

therefore tiebacks may be inclined at steeper angles to reach that strata faster. The selection of the angle of declination may simply be based on the desire of designer to pass through a job specific no-load zone as economically as possible.

Step 3. Calculate the anchor load from the horizontal strut load by modifying it for the angle of declination. In order to maintain good grout retention, the angle of declination is usually a minimum of 15 degrees.

Step 4. Based on known unit capacities of anchors in similar soils, design the length and diameter of the bond zone (that portion of the anchor which will provide the frictional resistance to the applied stress). A chart of typical values for various materials is appended in Chapter 18.4.

Note: There is considerable evidence that anchor capacity is not linear with diameter and length as this discussion would imply. However, the method outlined is a good first attempt and most anchors are designed in this manner. For specific situations involving very long anchors or problems with adjacent rights-of-way for anchor placement, more sophisticated methods may be necessary.

Step 5. Select the tendon size based on the maximum load which the anchor may carry. This load will be a function of the design load of the anchor but may be adjusted to account for the testing of the anchors (see the discussion in Chapter 14).

11.6 TOE CAPACITY (VERTICAL) OF SOLDIER PILES

There is no common understanding about the function of soldier pile toes when it comes to the tieback loads. Some believe that the vertical component of the tieback load is transmitted directly to the pile toe and therefore must be designed. Others believe that the vertical component of the tieback load is dissipated very quickly through friction to the lagging boards and the soil mass. Still others believe that the vertical load is dissipated, but only through the direct contact between the pile (in the case of driven soldier piles) or pile encasement (in the case of drilled piles) with the soil mass. There can be no question that vertical loading of pile toes and subsequent settlement of soldier piles from tieback vertical forces has occurred in cases where steeply sloping tiebacks were coupled with lagging which was not in tight contact with the soil mass behind it.

If the designer elects to design his soldier pile and lagging system such that all or most of the vertical component of the tiebacks is resisted by the toe, then it is necessary to sum the vertical loads and design the soldier pile toe as a drilled shaft. This may require the use of structural concrete and a deepening of the pile toe to provide adequate resistance. A similar analysis must be done for soldier piles retained by rakers (Chapter 11.7). A significant uplift load from the raker may require the design of the soldier pile toe as a tension drilled shaft.

11.7 RAKER FOOTINGS

Raker footings can be subdivided into two types: those footings which occur by bracing the raker against some portion of the new structure being constructed, and those footings which are constructed expressly for providing bearing capacity for the rakers.

Rakers which bear on some portion of the new structure are usually braced against the base slab of the structure. Given the raker load and the angle of placement of the raker, the horizontal and vertical forces being applied to the slab can be analyzed. In most cases, the mass of the slab is such that no specific adaptations must be made to the slab other than to define the method of attachment of the base of the raker to the slab.

Rakers which are designed to have their own footings are usually placed on footings which are excavated as deep narrow slots. Some rakers are attached to drilled shaft foundations (see Figure 4.11) or groups of driven piles, but most are founded on poured concrete footings (see Figure 11.9).

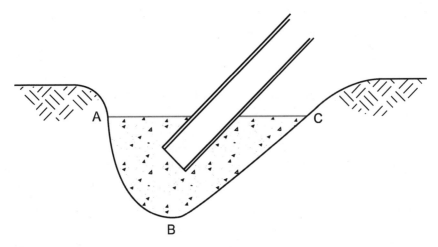

FIGURE 11.9   Typical raker footing.

The reasons for making the footings deep and narrow are many. Deeper and narrower means that there is less chance that the footing will interfere with the myriad of other installations in the base of the excavation such as building footings, sump pits, plumbing lines, etc. By making the footing deep, it is easier to mobilize significant passive pressure to resist the lateral load of the footing (Figure 11.10). In addition, deep narrow footings can be conveniently developed by pouring concrete against neat earth excavations without any forming. The footing is often only the width of the backhoe bucket.

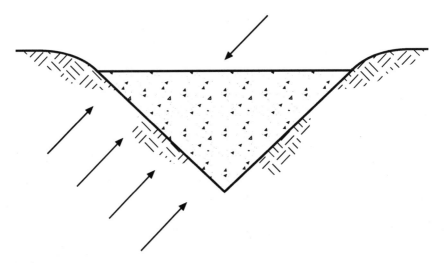

FIGURE 11.10   Raker load is restrained by inclined footing.

Using design methods outlined in Peck, Hanson & Thorburn for calculating the capacity of inclined footings, the size of the raker footing can be developed. In cohesive soils, the unit bearing capacity of the footing is defined as

$$q = cN_{cq} \qquad (11.9)$$

where q is the ultimate bearing strength
c is cohesion
N_{cq} is the bearing capacity factor (see Figure 11.11)

In cohesionless soils the ultimate bearing capacity of the footing is defined as

$$q = \tfrac{1}{2} B \gamma N_{\gamma q} \qquad (11.10)$$

where B is the inclined length of the footing bearing surface
γ is the unit weight of soil
$N_{\gamma q}$ is the bearing capacity factor (see Figure 11.11)

An alternative design method used to resist the lateral load placed on raker footings in softer soils involves the development of the capacity of the footing through adhesion between the sidewalls of the deep narrow concrete footing. Very large frictional areas exist which can carry significant load (Area ABC on Figure 11.9).

11.8 LAGGING

A large body of opinion holds that timber lagging should not be designed. This thought comes from observations that most lagging will simply deflect to the point where the retained soils will arch between the soldier piles and relieve the pressure on the lagging. Once a point of equilibrium is reached, it is argued, that deflection will stop.

Excavations of depths to 60 feet (18 m) with lagging thickness of 3 inches (75 mm) and spans of 10 feet (3 m) have performed well. Excavations to 110 feet (33.5 m) with 4 inch (100 mm) lagging and 9 foot (2.7 m) bays have similarly performed satisfactorily.

The designer should be cautioned that this principle does not hold in soft clays where arching is minimal or nonexistent. It should also be pointed out that in these types of materials, timber lagging, soldier pile and lagging is often not recommended at all.

That being said, there is a great desire on the part of many plan checkers to have some rational mathematical method of designing timber lagging. Goldberg Zoino in their report to the FHWA in 1976 (listed in the Bibliography) produced a chart of suggested lagging thicknesses which is accepted by some as sufficient for design purposes (see Table 11.1).

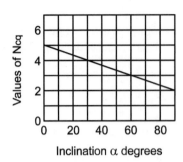

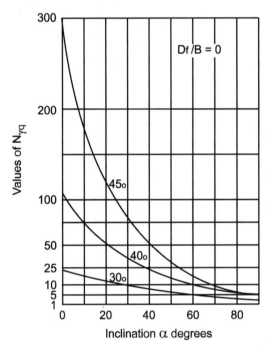

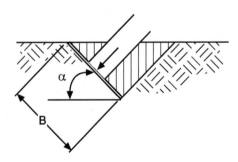

FIGURE 11.11 Raker footing design charts from Peck, Hanson & Thorburn.

TABLE 11.1 Goldberg Zoino Chart. (*Courtesy of the Federal Highway Administration*)

	Soil Description	Unified Classification	Depth	Recommended Thickness of Lagging (roughcut) for Clear Spans of:					
				5'	6'	7'	8'	9'	10'
COMPETENT SOILS	Silts or fine sand and silt above water table	ML, SM-ML							
	Sands and gravels (medium dense to dense).	GW, GP, GM, GC, SW, SP, SM	0' to 25'	2"	3"	3"	3"	4"	4"
	Clays (stiff to very stiff); non-fissured.	CL, CH	25' to 60'	3"	3"	3"	4"	4"	5"
	Clays, medium consistency and $\frac{\gamma H}{S_u} < 5$.	CL, CH							
DIFFICULT SOILS	Sands and silty sands, (loose).	SW, SP, SM							
	Clayey sands (medium dense to dense) below water table.	SC	0' to 25'	3"	3"	3"	4"	4"	5"
	Clays, heavily over-consolidated fissured.	CL, CH	25' to 60'	3"	3"	4"	4"	5"	5"
	Cohesionless silt or fine sand and silt below water table.	ML; SM-ML							
DANGEROUS SOILS	Soft clays $\frac{\gamma H}{S_u} > 5$.	CL, CH	0' to 15'	3"	3"	4"	5"	--	--
	Slightly plastic silts below water table.	ML	15' to 25'	3"	4"	5"	6"	--	--
	Clayey sands (loose), below water table.	SC	25' to 35'	4"	5"	6"	--	--	--

Note:
*In the category of "potentially dangerous soils", use of lagging is questionable.

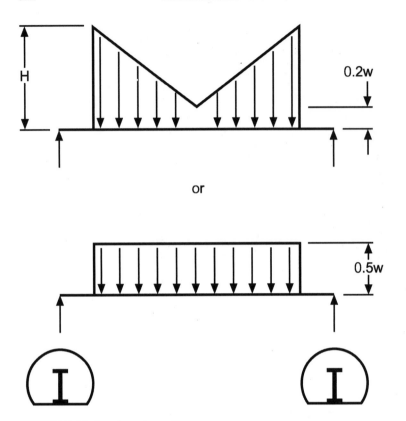

FIGURE 11.12 Lagging-pressure diagrams.

For those who continue to insist on a mathematical method, two loading diagrams are included as Figure 11.12 which are sometimes used. 'w' is the unit soil pressure from apparent earth pressure or earth pressure theory diagrams. The rationale behind these lagging diagrams is a follows. The unit pressure predicted by the active pressure diagram should be modified to account for the incidence of arching in the soils. No actual research is known to have been performed to create these pressure diagrams, but no failures of lagging boards are recorded by their use either.

11.9 SOIL NAILING

Soil nailing is always subject to some form of computer analysis. The following is a sample of the type of analysis which several of the recognized programs might follow.

Step 1. Determine the soil and dimensional parameters c, Φ, γ, q, and H.

Step 2. Determine the density of nails required to achieve continuity of the soil mass. This is usually assumed to be a 6 foot x 6 foot (1.8 x 1.8 m) pattern.

Step 3. Select a number of failure surfaces as trials (see Figure 11.13).

Step 4. For each failure surface, divide the soil mass into slices. Using the method of slices, determine the added force required to bring each slice into equilibrium.

Step 5. Determine the length of nail required for pullout resistance to provide the added normal force to create equilibrium. Use field experience or table in Chapter 18.4 to determine the length. Use the critical failure surface for each nail to determine the design load.

Note: The critical plane is not necessarily the same surface for each nail.

Step 6. Given the location of the critical slip surface for each nail, derive the length of nail by adding the length of embedment found in Step 5 (see Figure 11.14).

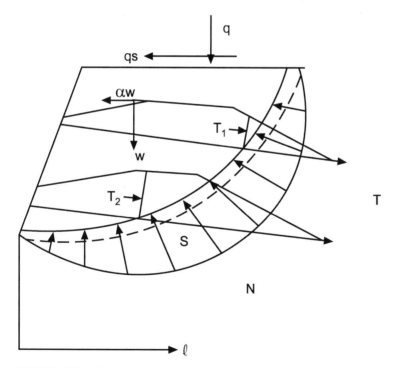

FIGURE 11.13 Method of slices. *(Courtesy Golder Associates, Inc., Redmond, WA)*

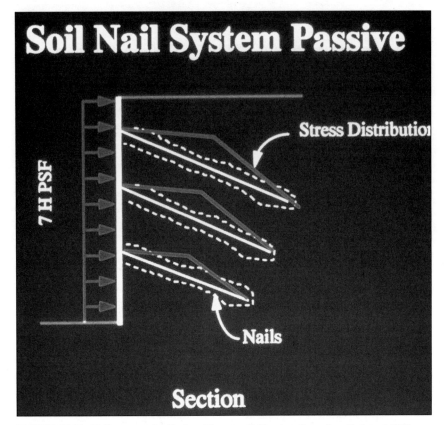

FIGURE 11.14 Nail pressure distribution. *(Courtesy Golder Associates, Inc., Redmond, WA)*

Step 7. Given the critical load in each nail, run trial slip surfaces between the critical surface and the wall face to derive the load distribution in nail. Dissipate the load from each slip surface through the adhesion assumed for the nail. This method will expose the amount of load which will ultimately be retained at the excavation face.

Note: Some designers have found that this step can be eliminated by assuming the fascia load is 30 percent of the nail design load.

Step 8. Design the nail head anchorage.

Step 9. Check the global stability of the system by applying an apparent earth pressure to the back of the resultant reinforced block (see Figure 4.5).

CHAPTER 12
GROUND WATER CONTROL

Probably no single issue causes as much disruption to an excavation project as does the presence of water. It can destabilize bearing surfaces, cause havoc with cut slopes, restrict the contractors choices when it comes to shoring, and cost time and money in efforts to deal with it.

Even when handled effectively, it is the primary cause of site access problems. The mud that is inevitable can turn one's neighbors into one's enemies, and can disrupt even the most meticulous of schedules. When not dealt with properly, it can have disastrous effects on all parties to the contract (see Figure 12.1).

When dealing with water problems on site, one must be prepared to deal with surface water, perched water, water tabled within the depth of the excavation as well as water pressures and aquifers below the depth of the excavation. In order to do so, a clear understanding of the types of water conditions to be encountered is necessary. This information must be combined with an evaluation of the potential effects of climatic events and seasonal variations. Only when this clear and rational picture of the issues is in place is it possible to properly design a water control plan which is essential to the successful excavation project.

FIGURE 12.1 Unwatering not well performed, Boise, ID. Note the ducks in the foreground. *(Courtesy of Condon-Johnson & Associates, Inc. Seattle, WA)*

12.1 UNWATERING

Unwatering is the term used to describe the process whereby water is removed from an excavation after it has entered the excavation. Many people confuse this procedure with dewatering, which it is not. Dewatering is the process used to prevent water from entering the excavation by actively pumping.

Unwatering may be the removal of water from inside a sheet pile cofferdam which is performed after the seal is placed in the base of the excavation. If the excavation is quite shallow (less than 15 feet (4.6 m)), the pumping is usually performed with vacuum pumps. Deeper excavations are handled with submersible pumps or trash pumps. If the water is relatively clean, it may be possible to return it to creeks, rivers or lakes either directly or by way of local storm sewers. However, water which is removed from an excavation usually carries an excessive silt load and must be treated before being returned to open water. This process may be as simple as broadcasting it over a large vegetated area and permitting the water to return to the ground water table through permeation.

If this is not possible, the water may need to be pumped to a settling basin which is formed by using either large tanks or lined pits dug for that purpose. Here the waterborne silts are allowed to settle and the clear water is decanted over a weir. In extreme cases, it maybe necessary to add a flocculant to encourage sedimentation or use desanding plants such as cyclones to remove solids. If the water contains specific pollutants it may need to be treated with either chemicals for precipitation or filtered to render it suitable for disposal.

Unwatering may also involve the removal of water which slowly accumulates in the low spots of an excavation and needs to be constantly removed. The source of this water could be rainfall, or perched water tables which drain into the excavation. This water is usually not of sufficient quantity to require dewatering and so is collected on site by a series of ditches which are constructed around critical elements of the work and drain towards one or several sumps for pumping and disposal (see Figures 12.2 and 12.3).

The question must be asked: "When should we dewater and when should we unwater?" The question is not only one of "what is possible" but also one of economics and schedule. If the water can be handled effectively by unwatering, the cost of the operation itself will almost always be cheaper than dewatering. However, the decision to unwater in lieu of dewatering can have far-reaching cost effects on shoring and excavation methods. The following are the types of tradeoffs that might need to be made:

- Unwatering may require the use of flatter side slopes for the excavation.
- Unwatering may restrict the types of shoring which can be used, i.e., it may require sheet piling instead of soldier pile and lagging or soil nailing.
- Unwatered sites will almost always be wetter and muddier and therefore more difficult to excavate than dewatered sites.

FIGURE 12.2 Unwatering trenches, Las Vegas, NV. *(Courtesy of Golder Associates Inc. Redmond, WA)*

FIGURE 12.3 Dewatering Sump, Seattle, WA. *(Courtesy of Malcolm Drilling Co., Inc. Kent, WA)*

In order to reduce the cost of unwatering, it is obvious that one should reduce the quantity of water pumped. This can be done in a variety of ways:

- Use localized sheeting or shoring when deep sumps or other localized excavations are required.
- Use water diverting methods as outlined in Chapter 12.3 to minimize the entry of water from outside the site.
- Consider the use of cutoff walls to minimize the water flow into the site from specifically identified water sources.

12.2 DEWATERING

Dewatering is the construction activity which is performed to remove ground water from a site prior to its entry into the excavation. Dewatering may be performed by drawing down the water table outside the site and maintaining this depressed groundwater level until the construction activities such as concrete work and grading, which must be performed below the ground water table have been completed. This type of general dewatering should only be performed when a thorough investigation of the possibilities of damage to adjacent property has been completed. General dewatering of this sort can affect neighboring wells, create settlement of adjacent properties, and reverse the natural flow of ground water which may cause the dispersion of pollutants already in the groundwater to areas which would not otherwise have been affected.

Dewatering that is carried on only within the boundaries of the construction site must be restricted by some form of barrier wall which will permit the ground water table outside the site to remain unaffected. These barrier walls could be sheet piling, secant walls, slurry walls, or could be slurry trenches filled with soil/bentonite, or soil/cement/bentonite mixtures. Once a barrier of this type is in place, the amount of water being pumped is quite reduced.

Dewatering can be performed by deep wells, wellpoint systems, or horizontal drains. Deep wells are almost always cheaper than wellpoints and are used when the native soils exhibit permeabilities which permit the creation of a broad drawdown curve. Wells are placed at 100 to 150 foot (30-45 m) centers and pumping from the wells is by high capacity submersible pumps. Wells maybe from 10 inches to 30 inches (250-760 mm) in diameter and have a screen and filter pack placed around them which permits the pumping of water without drawing in excessive sands and silts that might foul the pumping operation. This type of well performs best in sands and gravels down to permeabilities of 1.0×10^{-3} cm/sec. Large areas can be dewatered with a few wells (see Figure 12.4).

Alternatively, wellpoint systems can be installed. Wellpoints are small diameter wells which are either installed by drilling, driving, or jetting. The wells, which are screened over a discrete length near the base of the point are installed

FIGURE 12.4 Small sheeted excavation being dewatered by one deep well, Kent, WA.

at 2-8 foot (0.6-2.4 m) centers around the perimeter of the site requiring dewatering. The wells are attached to a central discharge line called a header. If the depth to be dewatered is less than about 15 feet (4.6 m), the wellpoints can be operated by vacuum and are called, quite logically, vacuum wellpoints. A vacuum is drawn on the header which then sucks the water out of the wellpoints. If depths of greater than 15 feet (4.6 m) are required, the use of vacuum wellpoints will necessitate the installation of another row of wellpoints approximately 15 feet (4.6 m) below the first in order to effect drawdown. This usually means that a sloped excavation profile is required although Figure 12.5 details a multi-level vacuum dewatering system which was installed through the face of a vertical soil nailed wall.

If dewatering is required to depths of greater than 15 feet (4.6 m), then the dewatering contractor may turn to eductor wellpoints. These wellpoints are also called ejectors. Each wellpoint operates by having a small quantity of water forced down the wellpoint under very high pressure. The return flow, traveling at lower pressure is able to lift ground water from the base of the wellpoint which has entered through its screen. The return flow is captured again in a header and directed to a disposal system similar to those discussed in Chapter 12.1

Wellpoints are effective in silts with permeabilities of around 1.0×10^{-5} cm/sec. Soils such as sandy silt, glacial silts and silty fine sands which fall in that intermediate range of permeabilities of 10^{-3} to 10^{-5} cm/sec may be dewatered by either method. It becomes a balancing act of reduced efficiency of deep wells or added costs of wellpoints.

Horizontal drains (Figure 12.6) are particularly effective when dealing with water which is flowing toward the excavation on top of a well defined strata. They are often used when a slide plane is being lubricated with groundwater and the overall stability of the slide mass can be improved by lowering the water table on the slide plane. These drains are installed by drilling horizontally into the face of the excavation and placing a slotted pipe protected by a filter fabric. Water is collected inside the excavation, piped to a pump location and disposed of. Horizontal drains of up to 700 feet (215 m) in length have been used to dewater and stabilize slide planes.

12.3 DIVERSION TECHNIQUES

In order to prevent the unwanted entry of surface water into the construction site with its attendant problems, contractors will often erect curbs or low check dams around the site or across locations of possible ground water entry. These can simply be raised concrete curbs or can take the form of eyebrow ditches (see Figure 12.7). These diversion structures direct water to a system of sumps for disposal of water to prevent its accumulation on site.

FIGURE 12.5 Two levels of vacuum wellpoints installed by drilling through the shoring face to provide dewatering prior to further excavation, Seattle, WA. *(Courtesy of Malcolm Drilling Co., Inc. Kent, WA)*

FIGURE 12.6 Horizontal drains drilled up to 700 feet (215 m) to drain a slide mass, Portland, OR. Note the collection system. (*Courtesy of Condon-Johnson & Associates, Inc. Seattle, WA*)

FIGURE 12.7 Curbing and eyebrow ditch installed prior to excavation, Portland, OR. *(Courtesy of Condon-Johnson & Associates, Inc. Seattle, WA)*

For water that cannot be diverted from a site such as rainwater, it is often a good idea to minimize its effects on cut slopes or penetration behind shoring systems. Water, when permitted to course down cut slopes, will destabilize the slopes by cutting erosion channels. If water is permitted to run behind the fascia of soil nail or lagging walls, it can cause instability of the entire system by scouring material from behind the wall and leaving voids.

The tops of all slopes adjacent excavations should have diversion curbing placed to prevent water from flowing down the slope. The top of the slope should be graded away from the site to further discourage the flow of water onto the site. The slopes, when subject to possible storm water deterioration, should be weather protected with either tarping or visqueen covering (see Figure 12.8). This cover should begin at the top of the slope and extend down to the top of the shoring and either form a lined ditch for collection of water or continue over the top of the shoring so the water is delivered into the excavation over the wall, not through the wall. It can then be collected on site for disposal through a series of sumps and ditches.

The consequences of permitting uncontrolled water flow on unprotected slopes is shown in Figure 12.9. Many contractors have found that the maintenance cost required to keep a tarped system in place is more expensive than the expense of protecting the slope with a thin layer of shotcrete (see Figure 12.10).

12.4 DRAINS AND COLLECTION

Despite the best laid plans, water will inevitably end up behind the shoring system. With systems which are not designed to withstand hydrostatic heads, this is of concern. These include soldier pile and lagging systems and soil nail systems. Soldier pile and lagging will usually dissipate any buildup of water pressure by leaking through the gaps in the lagging planks. This is perfectly acceptable as long as the flow of water does not bring fines with it. If it does, the loss of soil will eventually cause chimneying. If ground water flows in a specific lens of soil are found to be excessive, contractors will often stuff straw, or excelsior, behind the lagging to act as a filter to prevent the flow of soils.

Some designers will detail filter fabric and pea gravel filters behind lagging. These designs, while looking good on paper, are not constructible. The amount of over-excavation required to install these systems causes great disturbance behind the lagging. Subsequent lifts of lagging undermine this disturbed material and it inevitably falls out.

Once the lagging is complete, drainage fabric is often attached to the lagging. This fabric is then trapped between the lagging and the subsequent poured concrete wall and allows the downward flow of water from the wall to a footing drain. Once the water reaches the base of the wall in the drain fabric, it is piped through the wall and into a collection system for disposal.

FIGURE 12.8 Slope protection tarps, Renton, WA. (*Courtesy of Condon-Johnson & Associates, Inc. Seattle, WA*)

FIGURE 12.9 Erosion resulting from lack of slope protection, Vancouver, WA. *(Courtesy of Condon-Johnson & Associates, Inc. Seattle, WA)*

FIGURE 12.10 Slope protection utilizing shotcrete, Portland, OR. *(Courtesy of Condon-Johnson & Associates, Inc. Seattle, WA)*

With soil nailing, it is important that water does not build up behind the fascia. The fascia of soil nailing is shotcrete and is impermeable. In order to assure that water which gathers behind the fascia is dissipated, contractors install drain strips behind the shotcrete at approximately 6 foot (1.8 m) centers to move water to the base of the wall. The drain strips are installed in 6 foot (1.8 m) lifts which correspond to the shotcrete lifts. Each lift is cross communicated (see Figure 12.11) so that if a drain becomes blocked, the water will have an alternative flow path. Once the water reaches the bottom of the wall, it is piped through the shotcrete (Figure 12.12) with a gravity drain and pipe and collected in the building collection system (see Figure 12.13).

FIGURE 12.11 Drainage installed behind the shotcrete in a soil nail system, Beaverton. OR. *(Courtesy of Condon-Johnson & Associates, Inc. Seattle, WA)*

FIGURE 12.12 Through wall pipes drain accumulated water from drainage fabric, Seattle, WA. (*Courtesy of Condon-Johnson & Associates, Inc. Seattle, WA*)

FIGURE 12.13 Through wall pipes are plumbed into a collector system, Seattle, WA. *(Courtesy of Condon-Johnson & Associates, Inc. Seattle, WA)*

CHAPTER 13
INSTALLATION EQUIPMENT AND TECHNIQUES

The shoring industry is rife with inventors who develop equipment to deal with special situations, and it simply would not be possible to show all of the equipment involved. However, the following chapter will attempt to outline some of the equipment used for the installation and prosecution of shoring.

13.1 SHEET PILING

Sheet piling is driven by either vibratory or impact hammers. In some cases, the sheet is installed to refusal with a vibratory hammer and then finished off with an impact hammer. Vibratory hammers (Figure 13.1) are usually hung from conventional crawler cranes. Alternatively, sheeting is driven by vibratory methods with a sheeting driver which mounts a vibrating head on a fixed lead (Figure 13.2). This configuration permits crowd or pull down to be exerted together with vibration. Sheet piles can also be driven by lead mounted diesel or air hammers.

13.2. DRILLED PILES—DRILL AND PLACE

Included in this category are soldier piles, secant piles, cylinder piles and tangent piles. These piles are installed by drilling a hole and placing a steel section within the hole and then backfilling the hole with structural and/or lean mix concrete. Holes can be drilled with truck mounted drill rigs (Figure 13.3), crane mounted drill rigs (Figure 13.4), or crawler mounted drill rigs (Figures 13.5 and 13.6).

FIGURE 13.1 APE vibratory hammer used for pile driving—free suspended. *(Courtesy of Condon-Johnson & Associates, Inc. Seattle, WA)*

FIGURE 13.2 ABI vibratory hammer used for pile driving—lead mounted. *(Courtesy of ABI Inc. Benecia, CA)*

FIGURE 13.3 Watson truck mounted drilling machine. (*Courtesy of Watson, Inc. Ft. Worth, TX*)

FIGURE 13.4 Calweld crane mounted drilling machine. (*Courtesy of Deep Foundations Contractors. Thornhill, Ont.*)

FIGURE 13.5 Texoma crawler mounted drilling machine. *(Courtesy of Condon-Johnson & Associates, Inc. Seattle, WA)*

FIGURE 13.6 Soil-Mec crawler mounted drilling machine. (*Courtesy of Champion Equipment Co. Paramount, CA*)

Truck mounted rigs are extremely mobile and keep mobilization costs to a minimum. Truck mounts also have a good crowd system (down pressure applied to the Kelly bar). These rigs require excellent site conditions to permit movement and are being used less and less for the installation of soldier piles.

Crane mounted drill rigs are probably the most costly type of drill to mobilize to a site and require a great deal of skill in their use to accurately drill soldier pile holes. The distance from the driller's seat to the hole location places a special load on the driller when locating the hole. Crane mounts do have a large swing radius, which allows them to cast drill spoil over a wide area. These rigs tend not to become dirt-bound as quickly as other types of drill rigs. The rotary table is much higher than conventional drill rigs, and therefore they are very good for working with long casings when drilling conditions dictate that the casing must be advanced into the ground as drilling progresses.

Although they are available, most crane mounts do not have crowd systems and so are somewhat limited when drilling very hard formations. When drilling large diameter holes, such as cylinder piles, these rigs have a distinct advantage.

Crawler mounted drill rigs (Figures 13.5 through 13.8) are the most common form of drill rig used for pile drilling. The crawler mounting relieves the contractor of the excessive site development preparations necessary with truck rigs and yet they have good crowd systems, and are relatively cheap to move from site to site.

Configurations favored by American manufacturers of drill rigs (Figures 13.5 and 13.7) consist of a platform which houses the engine, transmission and some winches and pumps. The derrick, which can be lowered for shipping, rises from one end of the platform and the operator sits facing the platform and looks down on the hole he/she is drilling. Most drill rigs of this type feature the operator seated in an open air venue, although cabs for weather protection can be mounted (Figure 13.7). The rotary table is fixed at the base of the drill derrick. This type of drill rig is extremely accurate when drilling for plumbness as the distance from the tip of the mast to the rotary table is maximized which emphasizes verticality. The fixity of the rotary table however does limit the height of casing or tool that can be placed under the table.

The European form of this drill rig (Figure 13.6) consists of a crawler machine not unlike a trackhoe with a lead attached to the face of the machine. The lead can be lowered for shipping. The operator sits in an enclosed cab and must observe the hole from some distance behind the lead. The rotary table on these rigs is moveable and slides up and down the lead to permit the table to be raised to clear casing, handle long tools or to twist long casing (see Figure 13.8).

13.2.1 Drilled Piles—Low Head Room

Conventional drill rigs have derrick heights in the range of 60 to 100 feet (18-30 m) above ground. Their drill depths are from 40 to 160 feet (12-49 m). This is accomplished by nesting the Kelly bars (the drill steel which transmits the torque

FIGURE 13.7 Watson crawler mounted drilling machine. Note the operator cab. *(Courtesy of Watson, Inc. Ft. Worth, TX)*

of the drill platform to the drill head or auger) inside each other, so that the bars telescope out for added depth.

By nesting up to six elements together, manufacturers have developed rigs which can drill to depths of up to 90 feet (27 m) with mast heights of 27 feet (8.2 m) or less. Figures 13.9 and 13.10 are two such rigs. The rigs are mounted to trackhoes. As seen in Figure 13.9, the rig has now found favor in areas where added reach is needed.

FIGURE 13.8 Soil-Mec crawler mounted drilling machine. Note the variable elevation of rotary table. *(Courtesy of Champion Equipment Co. Paramount, CA)*

FIGURE 13.9 Bayshore Lo-Dril—low overhead drill. (*Courtesy of Condon-Johnson & Associates, Inc. Seattle, WA*)

FIGURE 13.10 Watson Excu-Dril–low overhead drill. *(Courtesy of Watson, Inc. Ft. Worth, TX)*

13.3. DRILLED PILES—WET SET

Piles may be placed in holes which are already prepared with soil cement or concrete. Auger cast rigs, such as the crane suspended model (Figure 13.11) or the lead mounted model (Figure 13.12), drill with a continuous flight auger and place concrete through the hollow stem of the auger. A soldier pile can then be wet set in the concrete, either by gravity, or by lightly vibrating the beam into the wet concrete.

Piles may also be installed into wet soil/cement. Figure 13.13 shows a Geojet rig mixing in-situ soils with cement. This process is known as the Deep Mixed Method (DMM). Figure 13.14 shows the mixing head which mixes high pressure cement grout while the rotation of the head mechanically breaks the soil's formation. Once the soil/cement column is prepared, the pile is lowered into the mixture, again either by gravity or vibration. Figure 13.15 shows the entire Geojet setup with the cement storage and grout mixing equipment.

13.4. PILES—DRIVEN

Soldier piles are also installed by driven methods. Vibro hammers similar to sheeting drivers (Figures 13.1 and 13.2), lead mounted diesel hammers (Figure 13.16), or drop hammers are used.

FIGURE 13.11 Auger cast drill rig, crane mounted. (*Courtesy of Hurlen, Inc. Seattle, WA*)

FIGURE 13.12 ABI auger cast drill rig, crawler mounted. (*Courtesy of ABI Inc. Benicia, CA*)

INSTALLATION EQUIPMENT AND TECHNIQUES **359**

FIGURE 13.13 Geojet soil mixing machine. *(Courtesy of Condon-Johnson & Associates, Inc. Oakland, CA)*

FIGURE 13.14 Geojet auger. *(Courtesy of Condon-Johnson & Associates, Inc. Oakland, CA)*

FIGURE 13.15 Geojet Rig with grouting equipment and cement storage. *(Courtesy of Condon-Johnson & Associates, Inc. Seattle, WA)*

FIGURE 13.16 Diesel pile driving hammer on fixed leads. (*Courtesy of Hurlen, Inc. Seattle, WA*)

13.5 TIEBACKS

Tiebacks may be drilled with any number of rigs (Figures 13.17 through 13.23). Auger cast tiebacks can be drilled with leads mounted on trackhoe frames (Figure 13.17) or crawler crane suspended (Figure 13.18). These rigs can also handle uncased auger or air rotary systems.

Conventional pile drilling rigs can be used for tieback drilling in situations where augers can progress tieback holes in ground suitable for excavation without casing (Figure 13.19).

Duplex drilling (the insertion of casing simultaneously with the drill rod) can be performed with rigs shown in Figure 13.20 or 13.21. The casing is usually placed in the chuck manually in 2 M lengths. In order to use longer lengths of casing, rigs have recently been introduced that have a carousel which will mechanically place the casing in the chuck (See Figure 13.21). At the time of publication it was still not clear which system is fastest.

In order to access difficult locations and even drill back under themselves, a duplex rig can have its drill mast dismounted and loaded onto either a trackhoe boom (Figure 13.22) or suspended on a crane platform (Figure 13.23).

13.6. TIEBACK GROUTING

Tieback grout is usually neat cement/water grout. In some instances, however, it consists of ready mix sand/cement grout and is placed by a conventional trailer mounted concrete pump (see Figure 13.24).

When mixing and pumping neat cement grouts, the contractor may use paddle mixers (Figure 13.25) or colloidal mixers (Figure 13.26). The hydration of cement is much more complete when using colloidal mixers and they seem to be gaining popularity. These two grout plants are usually fed with bagged cement.

When high volume grouting is necessary, contractors will often switch to hopper-fed plants which work with bulk cement (Figure 13.27). Mixing in these plants involves colloidal methods.

High pressure pumping for secondary grouting applications is performed by pumps which can either be mounted separately (Figure 13.28) or in tandem with colloidal or paddle mixers.

FIGURE 13.17 Bayshore "rocket launcher" tieback drill.

FIGURE 13.18 Crane suspended tieback drill. *(Courtesy of Condon-Johnson & Associates, Inc. Seattle, WA)*

FIGURE 13.19 Watson conventional auger rig drilling tiebacks. *(Courtesy of Watson, Inc. Ft. Worth, TX)*

FIGURE 13.20 Klemm dual rotary (casing and drill steel) top drive tieback drill. *(Courtesy of Condon-Johnson & Associates, Inc. Seattle, WA)*

FIGURE 13.21 Hutte dual rotary tieback drill with a casing carousel. (*Courtesy of Equipment Corporation of America. Philadelphia, PA*)

FIGURE 13.22 Trackhoe mounted Klemm dual rotary tieback drill. (*Courtesy of Condon-Johnson & Associates, Inc. Seattle, WA*)

FIGURE 13.23 Crane suspended Klemm dual rotary tieback drill. *(Courtesy of Condon-Johnson & Associates, Inc. Los Angeles, CA)*

FIGURE 13.24 Schwing trailer mounted concrete pump. (*Courtesy of Condon-Johnson & Associates, Inc. Seattle, WA*)

FIGURE 13.25 Chemgrout paddle mixer grout plant. (*Courtesy of Condon-Johnson & Associates, Inc. Seattle, WA*)

FIGURE 13.26 Chemgrout colloidal mixer grout plant. *(Courtesy of Condon-Johnson & Associates, Inc. Seattle, WA)*

FIGURE 13.27 Haney grout plant—colloidal mixer for high volumes. Note the cement hopper feeding the mixing pots. (*Courtesy of Condon-Johnson & Associates, Inc. Oakland, CA*)

FIGURE 13.28 Hi pressure grout plant for secondary grouting. (*Courtesy of Condon-Johnson & Associates, Inc. Seattle, WA*)

13.7 SLURRY WALLS

Slurry walls are generally excavated by bucket under slurry head (Figure 13.29). When extreme depths are necessary, or when digging is extremely tough, hydrofraise (Figure 13.30) are utilized which house cutter heads to pulverize the soil and rock into cuttings fine enough to be lifted to the ground surface with air lift principles (Figure 13.31).

13.8. SERVICE CRANES

Pile handling as well as service work required for stressing and material movement is often handled by crawler mounted conventional service cranes (Figure 13.32) Alternatively, rubber tired hydraulic cranes are used (Figure 13.33). In order to deal with difficult site conditions, contractors are now finding applications for crawler mounted hydraulic cranes such as that displayed in Figure 13.34.

13.9 EXCAVATION

The whole purpose of installing shoring is so that the soils and rock within the proposed excavation can be removed safely. A variety of equipment is used. Excavations are usually made with trackhoes of capacity from ¾ to 2 CY (0. 57-1.53 M^3). Figure 13.35 shows two smaller types of these machines. Loaders can be used on large sites (Figure 13.36) but should not be used for lagging excavation (See discussion in Chapter 5).

Excavated materials are loaded into dump trucks which can be tandem axle trucks (Figure 13.37), tractor trailer arrangements (Figure 13.38), or truck and pup combinations (Figure 13.39).

13.10 CONVEYORS

When excavation depths get to the point where it is not effective to put trucks into the excavation, conveyor systems are used. Figure 13.40 is a picture of a belt system which is mounted on the soil nailed wall of a 75 foot (23 m) deep excavation. Figure 13.41 is a picture of an elevating system which utilizes buckets instead of belts. This system permits operation at steeper angles than belt conveyors.

Figure 13.42 details a belt conveyor system which gains height by mounting to series of pile bents. Both belt and bucket conveyor systems are fed through a loading hopper detailed in Figure 13.43.

FIGURE 13.29 Soil-Mec slurry wall digging bucket. *(Courtesy of Champion Equipment Co. Paramount, CA)*

FIGURE 13.30 Soil-Mec slurry wall hydrofraise. *(Courtesy of Champion Equipment Co. Paramount, CA)*

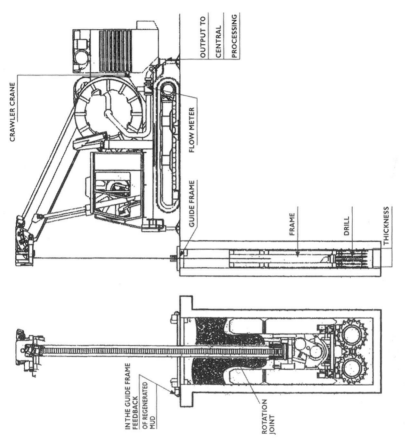

FIGURE 13.31 Hydrofraise schematic. (*Courtesy of Champion Equipment Co. Paramount, CA*)

FIGURE 13.32 Manitowac conventional crawler crane for drill service work. *(Courtesy of Condon-Johnson & Associates, Inc. Seattle, WA)*

FIGURE 13.33 LinkBelt rubber tired hydraulic service crane. *(Courtesy of Condon-Johnson & Associates, Inc. Seattle, WA)*

FIGURE 13.34 Mantis crane for drill service work. (*Courtesy of Condon-Johnson & Associates, Inc. Seattle, WA*)

FIGURE 13.35 Kobelco crawler track excavator. (*Courtesy of City Transfer Inc. Kent, WA*)

FIGURE 13.36 Caterpillar rubber tired loader. *(Courtesy of City Transfer Inc. Kent, WA)*

FIGURE 13.37 Tandem axle dump truck. (*Courtesy of City Transfer Inc. Kent, WA*)

FIGURE 13.38 Trailer dump truck. *(Courtesy of City Transfer Inc. Kent, WA)*

FIGURE 13.39 Truck and pup combination dump truck. *(Courtesy of City Transfer Inc. Kent, WA)*

FIGURE 13.40 Wall mounted conveyor system for moving and loading excavated materials. (*Courtesy of Golder Associates Inc. Redmond, WA*)

FIGURE 13.41 Bucket conveyor for moving and loading excavated materials. *(Courtesy of Fruhling Excavating, Inc. Seattle, WA)*

FIGURE 13.42 Pile supported conveyor system for moving and loading excavated materials. *(Courtesy of City Transfer Inc. Kent, WA)*

FIGURE 13.43 Loading hopper for conveyor systems. (*Courtesy of City Transfer Inc. Kent, WA*)

CHAPTER 14
LOAD TESTING OF ANCHORS

Anchored earth retention is probably one of the most tightly tested and confirmed systems available in geotechnical engineering. After the geotechnical engineer has characterized the soils and the designer has selected and designed the shoring system, the anchoring is subjected to an intense battery of tests. The design assumptions for the anchor length and installation method are tested through a series of verification tests. Once the design assumptions are confirmed, the actual installation of the anchors is rigorously tested. Through a system of performance tests, confirmation is obtained that the agreed assumptions are still in force. Then, regular proof tests are performed on the remainder of the anchors to ensure that quality is being maintained. In the end, every anchor on the project should have been tested in one fashion or another (Figure 14.1).

14.1 VERIFICATION TESTS

In order to begin soil or rock anchor design, an initial assumption of capacity must be made. This assumption is then tested through a process called verification testing which is used to test the soil/grout interface. The verification test is not used to test the grout/ tendon interface because it is well understood and not subject to job specific differences. Similarly, the verification does not test the tendon strength. Extensive regular material testing by manufacturers of anchor tendons and their components is performed prior to the sale of the tendons. Verification anchors are usually installed as sacrificial anchors prior to the start of construction. These anchors are ideally loaded to failure. By failing the anchors, the engineer has an

FIGURE 14.1 Test stressing of strand tiebacks. *(Courtesy of Deep Foundations Contractors, Thornhill, Ont.)*

accurate understanding of the ultimate capacity of the grout/soil interface. In order to do so, it is important that the test tendon itself be designed so that it is strong enough not to fail before the soil/grout interface.

Verification tests are usually installed and stressed against some form of grillage placed on the ground (see Figure 14.2). The movements of the anchor are recorded by measuring deflections of the anchor head with an independent measuring frame. Note that Figure 14.3 has a system of dial gauges set up on a tripod. Dial gauges are capable of measuring movements in increments of 0.001 inch (0.025 mm). It is important that the dial gauge measuring system be set up independently of the jacking frame (soldier piles or grillage) since any attachment of the gauges to the jacking system will result in the measurement of not only the anchor movement but also the settlement of the jacking frame.

Anchor stressing is performed utilizing hydraulic rams. Hydraulic pressure, read on a gauge, is used to calculate the total load being applied. The gauge and ram must be calibrated by a licensed testing facility so that an accurate chart of readings vs. actual ram forces is created. Figure 14.4 is a sample of one such calibration chart.

A testing program should be designed with the expectation that the anchor will demonstrate a capacity of at least two times (2x) the design load. The test load is usually applied in increments of 25 percent of the proposed design load. An initial load of approximately 10 percent of the design load is applied to the anchor which allows the anchor and jack to align themselves and work any slack out of the grillage. Once the alignment load is in place, measurements of all further loads and movements are recorded. A typical loading cycle for verification testing is listed herein. Some agencies specify that loading is cycled on and off in the following manner.

AL (Alignment Load)
0.25 DL (Design Load)
AL
0.25 DL
0.50 DL
AL
0.25 DL
0.50 DL
0.75 DL
1.0 DL
AL
0.25 DL
0.50 DL
0.75 DL
1.0 DL

(continued on page 399)

FIGURE 14.2 Verification test set against a grillage. (*Courtesy of Condon-Johnson & Associates, Inc. Seattle, WA*)

FIGURE 14.3 Dial gauges on a tripod. *(Courtesy of ADSC-The International Association of Foundation Drilling, Dallas, TX)*

CHART FOR LBS, PSI READING

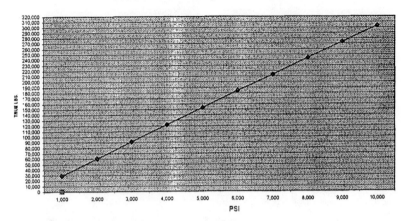

TEST INST., PSI	LBS., FORCE	CALCULATED LOAD	% OF ERROR	KPA	KN
1,000	29,087	29,370	-0.97%	6,895	134.37
2,000	60,416	61,000	-0.97%	13,790	268.74
3,000	91,450	91,500	-0.05%	20,685	402.33
4,000	122,248	122,000	0.20%	27,580	533.02
5,000	153,105	152,500	0.40%	34,475	673.96
6,000	183,903	183,000	0.49%	41,370	808.33
7,000	213,226	213,500	-0.13%	48,265	941.91
8,000	244,142	244,000	0.06%	55,160	1,074.97
9,000	273,524	274,500	-0.36%	62,055	1,213.28
10,000	303,260	305,000	-0.57%	68,950	1,358.15

LINEAR RELATIONSHIP:
FORCE (LB.)=FACTOR X PRESSURE(PSI)
FORCE (N)=FORCE (LB) X 4.4482
(R^2=0.9997)
PSI X 0689=KPA
0 TO 1,999 PSI, FACTOR IS 29.37
2,000 TO 10,000 PSI, FACTOR IS 30.5

Test Equipment Used: Revere Load Cell 500k
Serial No.: 890695UL
NIST Traceable No.: SCC-8455
Date Calibrated: 5/24/00
Service Engineer: MARSHALL DOYLE

FIGURE 14.4 Jack calibration chart. *(Courtesy of Condon–Johnson & Associates, Inc. Seattle, WA)*

(continued from page 395)

1.25 DL
1.50 DL
1.75 DL
2.00 DL
unload

while others will merely load incrementally as follows:

AL
0.25 DL
0.50 DL
0.75 DL
1.00 DL
1.25 DL
1.50 DL
1.75 DL
2.00 DL
unload

For a typical stress test record form (see Figure 14.5). At each load increment the load can be held for a period of about one minute to ensure that movements in the anchor have ceased prior to moving on to the next load increment. Other specifiers may demand a 10 minute load hold at each increment. In order to assure that the load is being maintained at a constant level, verification testing usually includes a load cell (see Figure 14.6). The load cell is also a calibrated instrument which should be confirmed by an approved testing laboratory in a manner similar to the ram and gauge.

The primary purpose of a load cell in a verification testing program is to identify changes to the applied load during the load hold period. These changes may not be evident by reading the ram gauge only since there may be some internal ram friction that masks the changes in applied load.

Huge arguments can occur when inspectors attempt to correlate load cell readings with ram gauge readings. It is very difficult to maintain the entire system in absolute agreement, so the following protocol is usually exercised. The initial load in any increment is established by using the ram gauge. An immediate reading of the load cell is taken. Any change in the load cell during the testing of a particular increment of load should be corrected by adjusting the ram pressure to keep the load cell constant. So, the load increments are set using the ram gauge, and constancy is maintained using the load cell.

In some cases, when the anchor reaches the design load (1.00 DL), a creep test is run. Others will run their creep tests at test load (2.00 DL). This test is an extended hold to monitor the load holding capacity of the anchor. With temporary anchors, the creep test is usually about 60 minutes provided that satisfactory

Project Name:	B. Clark/Faelnar Estates Apartments/WA	Tieback Designation:	T2
Project Number:	013-2694-103.009		
Contractor:	Cabe Construction, Inc.	Installed Tendon Length (ft):	58.0
Inspector:	Eric Lindquist	Installed Unbonded Length (ft):	33.0
Stressor:	Colin Cox	Bond Length (ft):	25.0
Date:	06/25/2000	Test Unbonded Length (ft):	33.0
Ram Serial Number:	WIKA 5458		
Gauge Number:	WIKA 5458	Tendon Grade (ksi):	270
		Allowable Stress Coefficient:	0.80
		Tendon Size:	9@0.6"DIA
Calibration Equation:		Tendon Area (in):	1.740
	Pressure (psi) = M [Load (k)] + C (psi)	Design Adhesion (k/ft):	3.50
	M = 12.0 (psi/k)	Design Load (k):	87.5
	C = 0.0 (psi)	Test Type (Proof, Verif):	VERIF

Time	Loading			Movement				
	Load Increment (%)	Load Increment (k)	Gauge Pressure (psi)	Gauge 1 (in)	Gauge 2 (in)	Average (in)	Comments	
	AL	-	-					
(0)	25	21.9	262	-	-	-		
(10)	25	21.9	262	-	-	-	Allowable Bar Load (k):	375.8
(0)	50	43.8	523	0.196	0.103	0.1495		
(1)	50	43.8	523	0.196	0.103	0.1495		
(0)	75	65.6	785	0.413	0.26	0.3365		
(1)	75	65.6	785	0.413	0.26	0.3365		
(0)	100	87.5	1047	0.572	0.491	0.5315		
(1)	100	87.5	1047	0.572	0.491	0.5315		
(0)	125	109.4	1308	0.729	0.526	0.6275		
(1)	125	109.4	1308	0.729	0.526	0.6275		
(0)	150	131.3	1570	0.929	0.713	0.821		
(1)	150	131.3	1570	0.929	0.712	0.8205		
(2)	150	131.3	1570	0.929	0.711	0.82		
(3)	150	131.3	1570	0.932	0.710	0.821		
(5)	150	131.3	1570	0.932	0.710	0.821		
(6)	150	131.3	1570	0.932	0.710	0.821		
(10)	150	131.3	1570	0.932	0.710	0.821		
(20)	150	131.3	1570				STOPPED AT 10 MIN.	
(30)	150	131.3	1570					
(50)	150	131.3	1570					
(60)	150	131.3	1570					
(0)	175	153.1	1832	1.158	0.923	1.0405		
(1)	175	153.1	1832	1.158	0.923	1.0405		
(0)	200	175.0	2093	1.402	1.151	1.2765		
(1)	200	175.0	2093	1.4025	1.1505	1.2765		
(0)	225	196.9	2355	1.771	1.508	1.6395	PRESSURE @2300 PSI	
(1)	225	196.9	2355	1.771	1.509	1.6400		
(0)	250	218.8	2616	-	-	-	DID NOT REACH	
		209.0	2500	-	-	-	MAX. PRESSURE BEFORE FAILURE	
		200.7	2400	2.304	2.12	2.2120	DIAL READING JUST AFTER FAILURE	

FIGURE 14.5 Tieback stressing record form. *(Courtesy Golder Associates Inc. Redmond, WA)*

results are obtained. With permanent anchors some tests may extend for five hours or even 24 hours in some cases. Measurements of deflection should be taken at the following intervals: 1, 2, 3, 4, 5, 6, 10, 15, 20, 25, 30, 45, 60 minutes and if specified 75, 90, 100, 120, 150, 180, 210, 240, 270, 300 minutes. In order to assure that any movement in an anchor is decreasing and will become negligible over time, the movements during a creep test are plotted against time. To assure this standard, the anchor movements must be held to under 0.08 (2 mm) inches per log cycle and be decreasing. This means that the total movement of the anchor under sustained load might be 0.08 inches (2 mm) in the first 10 minutes, and less than 0.08 inches (2 mm) in the next 100 minutes. Using this philosophy

FIGURE 14.6 Jack setup with load cell. *(Courtesy of ADSC-The International Association of Foundation. Drilling Dallas, TX)*

you can see that the next 0.08 inches (2 mm) would not occur in less than 1,000 minutes, and so on. Very quickly the anchor stops moving. Creep testing is of greater importance in cohesive soils (Plasticity Index of greater than 20) where the soil mass is subject to consolidation changes as the anchor load is applied. By checking that the movement is decreasing over time, the designer is assured that a slow creeping failure will not occur.

As the loads are being applied and the movements being measured, they are plotted to develop an understanding of the performance of the anchor. As discussed in Chapter 4.4, the anchor consists of a no load zone where no load is to be shed to the surrounding soil, and an anchor zone where the entire load is to be focused. Because steel deforms uniformly with stress, it is possible for the load test inspector to compare the theoretical movement of the anchor in the no-load zone with the measured movement of the anchor. The theoretical elongation of the tendon no-load zone can be calculated by the use of the following equation.

$$\Delta = \frac{PL}{AE} \qquad (14.1)$$

where

Δ is the elongation at load P (inches)
P is the total load on the anchor (Kips)
L is the length of the no-load zone (inches)
A is the area of steel of the tendon (square inches)
E is the Young's modulus of steel (30,000 KSI)

If the movement of the anchor is less than the theoretical elongation of the no load zone, it is evidence that the no-load zone is not functioning properly and that load is being lost in the no-load zone. This means that the anchor zone is not being subjected to its intended test. When elongation of an anchor is less than 80 percent of the theoretical elongation of the no-load portion of the anchor, the anchor is usually rejected.

If the verification tests indicate that the anchor did not fail at 200 percent of design load and that its creep performance is satisfactory, then the construction of the shoring system can progress without changes to the shoring design. If the test is taken to failure (at either less than or greater than 200 percent DL), then the anchor design can be changed so that the anchor adhesion values at design load will represent 50 percent of the failure values.

Once a successful verification test program has been performed, the construction procedures used for the production anchors must duplicate the verification test. The drilling and grouting equipment should not change in type, the grouting procedures should duplicate those used in the test, and the anchor grout mixes should not be changed. The only change which would not affect the validity of the verification test would be an alteration to the amount of steel provided in the anchor tendon. Verification anchors are usually built with added steel to allow for increased stressing. Anchors should never be tested to capacities greater than 80 percent of Guaranteed Ultimate Test Strength (GUTS). All personnel involved in a testing program should ensure that this concept is maintained. The amount of force contained in a stressing test can be extremely dangerous if suddenly released by the breaking of the anchor tendon.

Verification testing of soil nails is carried out in much the same fashion as soil anchors. Soil nail verification tests usually do not have a significant no-load zone (probably 3-10 feet (0.9-3.0 M)) and the no-load zone in a soil nail is left totally vacant. Because of this, the necessity to check the elongations of soil nails during verification testing is not as important as with anchors.

14.2 PERFORMANCE TESTS

Performance tests are carried out on a representative number of anchors or nails during production. Five percent of anchors are often tested. Performance tests involve loading to 1.33 times design load (sometimes 1.5 times). A typical performance test might be:

AL (Alignment Load)
0.25 DL (Design Load)
AL
0.25 DL
0.50 DL
AL
0.25 DL
0.50 DL
0.75 DL
AL
0.25 DL
0.50 DL
0.75 DL
1.00 DL   hold for creep test
1.25 DL
1.33 DL
1.00 DL   lock off (Note: no lock off for soil nails)

Other variations of performance tests may involve running the load up in increments without cycling. Still other test procedures may involve creep testing at the highest test load (1.33 DL or 1.5 DL) instead of 1.00 DL.

A performance test can be performed on any production soil or rock anchor which is selected by the inspector. Because soil nails are designed to hold the maximum load in the middle of the nail (see Figures 11.13 and 11.14), it is necessary for the inspector to designate the soil nail to be performance tested prior to installation. This will permit the contractor to install a no-load zone of sufficient length that the nail is not overstressed, while still testing the anchor adhesion to limits sufficient for satisfaction of the performance test criteria. As with verification tests, the maximum test load must never be taken above 80 percent GUTS.

The test load at 1.00 DL (or maximum load if so specified) is held for ten minutes. Deflection measurements should be made at the following intervals: 1, 2, 3, 4, 5, 6, 10 minutes. Creep criteria used for judging performance tests should be as follows. If the creep in the first 10 minutes is less than 0.04 inches (1 mm), then the anchor is deemed to have passed the creep test. If the elongation is greater than 0.04 inches (1 mm), the load is be held for an additional 50 minutes and the movement readings recorded at 20, 30, 40, 50 and 60 minutes. If the creep rate in the period 6 minutes to 60 minutes does not exceed 0.08 inches (2 mm) and is decreasing, the anchor is considered to be acceptable.

Measurements are plotted on a log-scale. By plotting the elongation vs. load, the inspector can ensure that the no-load zone is functioning. See the discussion of elongation in Chapter 14.1. Performance tests do not always utilize load cells. This seems to be left to the discretion of the individual specifier.

14.3 PROOF TESTS

Proof tests are performed on all anchors not otherwise tested. Soil nails are not proof tested. A typical test procedure is as follows:

AL
0.25 DL
0.50 DL
0.75 DL
1.00 DL
1.25 DL
1.33 DL   10 minute hold
1.00 DL   lock off

    Measurements are taken similar to those for verification tests. Plotting of all measurements is the same as used for verification tests. Creep tests use the same acceptance criteria as performance tests and the elongation of the no-load zone is checked for conformance with design expectations in the same manner as verification and performance tests. Load cells are not used for proof tests.

    It is not necessary to lock off tiebacks at 100 percent of their design load. In fact, in the past it was quite common to lock in something less than the full design load. However, recently the norm seems to be for engineers to specify 1.00 DL as the lock off load.

14.4 PLOTS

Figure 14.7 is a plot of creep versus log time used to determine creep acceptability of anchors. As discussed earlier (Chapter 14.1), in order for the anchor to be acceptable at the proposed load, creep must be less than 0.08 inches (2mm) per log cycle and decreasing.

    Figure 14.8 is a plot of load versus elongation used to check ultimate capacity of the anchor as well as the length of the no-load zone of the anchor. The ultimate anchor capacity is reached when the load remains constant or decreases as the elongation increase. You will note that the test log has two pre-plotted lines. The A line is the calculated elongation of the no-load zone only (use Equation 14.1). Chapter 14.1 discussed the rejection of anchors which elongate less than

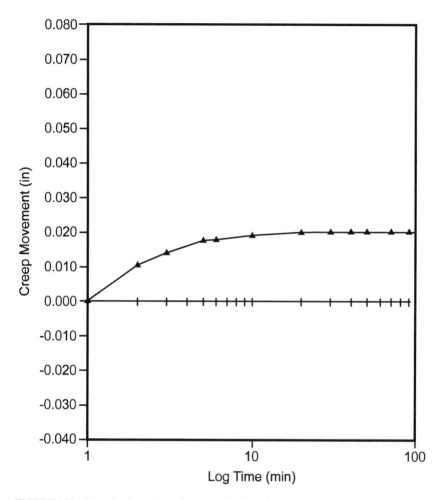

FIGURE 14.7 Creep testing—elongation versus log-time plot.

80 percent of that theoretical length. The first line shown in the plot is called 0.8A and is, in fact, 80 percent of the theoretical elongation of the no load zone. If the test curve does not end up right of the 0.8A line, then a problem is occurring in the no-load zone and the anchor should be rejected. The second line called out on Figure 14.8 is the B line. It is the theoretical elongation which would occur if the total load were applied to the no-load zone plus 50 percent of the anchor zone. If the test curve moves right of the B Line, then considerable movement is occurring in the anchor zone. While it is possible that this could occur in an acceptable anchor, this result should be reviewed by the geotechnical engineer prior to acceptance of the anchor.

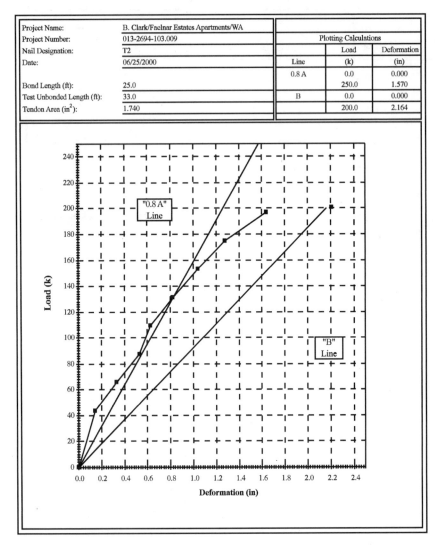

FIGURE 14.8 Tieback load versus elongation plot. *(Courtesy Golder Associates Inc. Redmond, WA)*

CHAPTER 15
MONITORING

An often overlooked part of any shored excavation is the monitoring of movements and forces in the shoring system during excavation. Without a systematic method of monitoring loads and movements, the contract team is without any early warning system to deal with unexpected occurrences.

Since the design of any shoring system is an exercise in estimation and prediction, it is extremely important to verify the pre-construction assumptions. The soils are modeled as best we can, but any modeling is necessarily a simplification. No design method currently in use can accurately predict ground movements. The only deflection calculation methods used are based on experience and comparison of similar cases.

If a carefully planned system of excavation monitoring is undertaken, movements which are observed can be analysed for conformity with the design intent. These movements are often quite slow in their accumulation and changes to the excavation sequence or structural modifications can be undertaken which will limit movements prior to the onset of serious damage to adjacent facilities.

Movement of adjacent facilities as well as the shoring system can be monitored by surveying techniques. Movement of the shoring system can also be measured very accurately by a system known as slope indicator readings. Forces in structural members can be determined by strain gauge measurements and can be back calculated from measurements of pile movements.

15.1 SLOPE INDICATOR MEASUREMENTS

Invented in 1958 by Shannon & Wilson of Seattle, WA, the slope indicator system utilizes extremely precise gyroscopic measuring devices to determine the verticality of the instrument in which they are housed (see Figure 15.1). Prior to excavation, a slope indicator casing is installed either in a pre-drilled hole, or attached to the piling used for shoring. The casing has several tracks inscribed into its interior which act as a guide for the slope indicator. The casing is installed to a depth which should be below any ground movement.

The slope indicator is lowered into the casing and measurements of verticality of the instrument are made at regular intervals. A plot of the shape of the casing can then be made from these readings (see Figures 15.2 and 15.3). Readings are taken at regular intervals during the excavation and the variations in the shape of the casing are indicative of any movements which may be occurring in the retained soil mass. Because the bottom of the casing is assumed not to move, it is possible to use the tip of the casing as a reference point and plot the changing shape of the casing as horizontal movements to an accuracy in thousandths of inches (0.025mm).

It is possible to monitor shoring movements and verify that the stressing of tiebacks or the preloading of struts is having the desired effect. Readings are taken every week during periods of little or no excavation, and as often as every day, during periods of excavation and stressing. If movements are noted which are of concern, readings should be taken at daily intervals until resolution of the concern occurs.

Slope indicators should always be read in concert with a series of pile survey readings to compare the readings and confirm the assumptions being made in the plotting of the slope indicator readings. By comparing the top of the pile location from survey data with its predicted position from the slope indicator readings, it is possible to confirm that the bottom of the casing has not moved. Figure 15.2 details a series of slope indicator readings showing the movements caused by excavation with the level of excavation indicated on each day of reading. Figure 15.3 details movements on successive readings over a period of 22 days. In this case, the information is not provided to conclude whether the movements are the result of excavation.

Slope indicator readings are usually not undertaken in excavations which do not have sensitive facilities adjacent to them. Slope indicator measurement methods are most often seen on deep excavations of greater than 35 feet (10.7 m).

15.2 PILE MOVEMENTS

Probably the easiest monitoring measurements taken for a shored excavation are performed with survey instruments. A baseline is established along the face of a shored wall and monuments are placed on the wall. In the case of soldier pile and

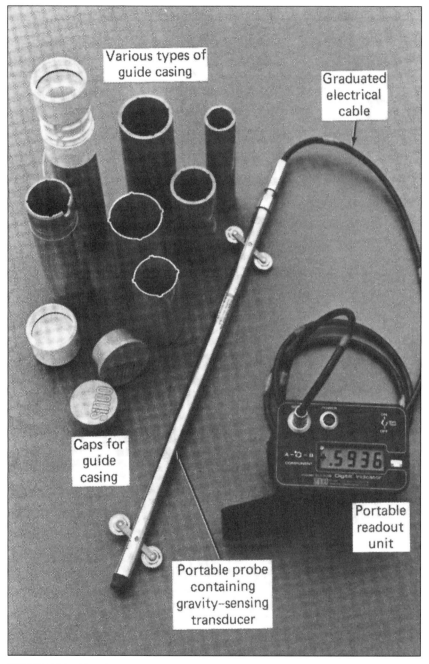

FIGURE 15.1 Slope indicator equipment. *(Courtesy of Isherwood Associates. Oakville, Ont.)*

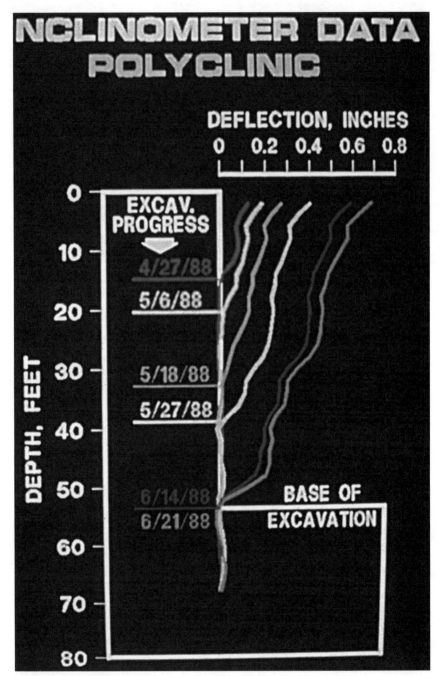

FIGURE 15.2 Slope indicator readouts from soil nailed excavation. *(Courtesy of Golder Associates Inc. Redmond, WA)*

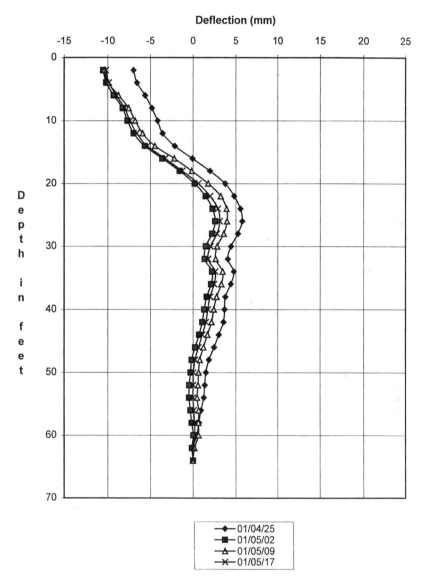

FIGURE 15.3 Slope indicator readouts from soldier pile and lagging shored excavation. *(Courtesy of Isherwood Associates. Oakville, Ont.)*

lagging each pile is monitored. In the case of sheet piling or slurry walls, monuments are established at regular intervals (5-10 feet (1.5-3.0 m)). It is important that the survey baseline is tied to monuments which are well outside the zone of any possible ground movement and is reconciled through a series of global positioning checks. Prior to excavation, a set of readings must be taken to establish the location of each monument or pile prior to any potential ground movement.

Regular readings are taken during excavation and stressing operations to identify any movements that may be occurring. These readings should be taken with the same regularity as the previously outlined slope indicator readings.

Readings can be undertaken with a survey transit. Welded flat bar can be placed at each measurement location with premounted scales attached to each (Figure 15.4). It is then possible for measurements to be taken by one man. If premounted scales are not used, it is necessary for someone to hold a scale at each pile so that the instrument man can take the reading. It is important that the exact location of the measurement point be marked on the pile in order that continuity of measurements is maintained.

When performed with diligence, readings to an accuracy of $\frac{1}{32}$ inch (1 mm) are possible. Readings should be reported as a series of contours (see Figure 15.5). Although the location of the piles will not be precisely in a straight line due to installation tolerances, the contours should be zeroed to indicate net movement along the wall.

15.3 ADJACENT STRUCTURE MONITORING

Any structure thought to be within the limits of any anticipated ground movements should be monitored. Benchmarks should be established on these structures to monitor vertical settlement. Any cracks in the adjacent structures which are evident from the pre-construction survey (see Chapter 7.2) should be covered with crack telltales (see Figure 15.6).

Observations of these telltales should be undertaken on a daily basis if any activity is ongoing in the excavation, or if any movements are reported in the shoring monitoring system. If no movements are evident in the monitoring system, and excavation has ceased, observations should be made at least weekly until the excavation is backfilled.

15.4 STRAIN GAUGES

Forces which are developed in rakers, walers, or struts can be monitored by the placement of strain gauges on the elements. Because these elements are almost always steel and the stress strain relationship for steel is well known, any strain which is measured in a steel member can be converted into a stress. The stress can then be compared to predicted stresses from the design calculations. If variations are found which are of concern, changes can be made to alleviate the problem.

FIGURE 15.4 Scales mounted on welded flatbar for easy reading of pile deflection. Installation on secant pile wall, Toronto, Ont. *(Courtesy of Isherwood Associates. Oakville, Ont.)*

FIGURE 15.4 *(continued)* Installation on soldier pile wall, Toronto, Ont. *(Courtesy of Isherwood Associates. Oakville, Ont.)*

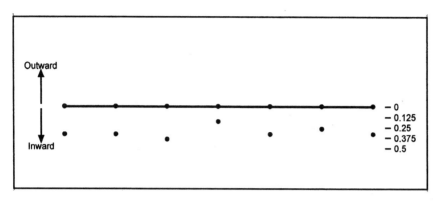

FIGURE 15.5 Typical pile movement plot from survey data.

Vibrating wire strain gauges are produced by a number of geotechnical instrument companies. These strain gauges are epoxied or welded to the structural member and strain at the same rate as the member. An electrical current is passed through the strain gauge and variations in its resistance are transcribed into strain measurements (see Figure 15.7).

This method is most often installed on struts in deep excavations such as subway cuts. Measurements should be taken daily during any period of excavation and continued if any unwarranted buildup of stresses is noted. Measurement in the range of 1/10,000 inch (0.0025 mm) are possible which can then be converted into steel stresses which are accurate to within 100 psi (0.7 MPa).

Strain gauge measurements should be compared to pile movement measurements in order to confirm the veracity of both systems. Monitoring readings should be transcribed into a regular report and distributed to the owner, general contractor, design engineer, and specialty subcontractor responsible for the shoring within 24 hours.

TELL-TALES consist of two plates which overlap for part of their length. One plate is calibrated in millimeters and the overlapping plate is transparent and marked with a hairline cursor. As the crack width opens or closes, one plate moves relative to the other. The relationship of the cursor to the scale represents the amount of movement occurring.
Range ±20mm – Resolution 1mm

STANDARD

The Standard TELL-TALE is produced in durable acrylic plastic and is used for monitoring movement across cracks in vertical and horizontal directions.
Re-order Code TT1
Linear coefficient of thermal expansion: $6.5 cm/cm°C \times 10^{-5}$

CORNER

The Corner TELL-TALE is produced in PVC rigid sheet and is used for monitoring movement across cracks in corners in vertical and horizontal directions.
Re-order Code TT2
Linear coefficient of thermal expansion: $6.5 cm/cm°C \times 10^{-5}$

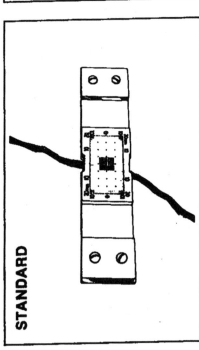

FIGURE 15.6 Telltales for indicating crack propagation. (*Courtesy of Shannon & Wilson, Inc. Seattle, WA*)

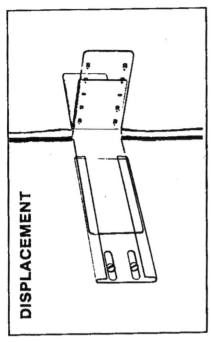 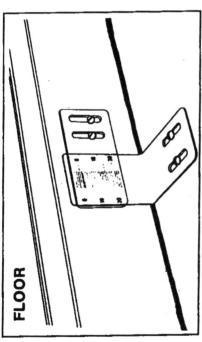 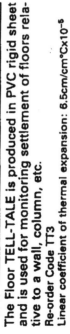

The Floor TELL-TALE is produced in PVC rigid sheet and is used for monitoring settlement of floors relative to a wall, column, etc.
Re-order Code TT3
Linear coefficient of thermal expansion: $6.5 \text{cm/cm}°\text{C} \times 10^{-5}$

The Displacement TELL-TALE is produced in PVC rigid sheet and is used for monitoring movement across cracks when one surface moves out of plane with the other. The TELL-TALE monitors out of plane movement.
Re-order Code TT4
Linear coefficient of thermal expansion: $7.3 \text{cm/cm}°\text{C} \times 10^{-5}$

FIGURE 15.6 *(continued)* Telltales for indicating crack propagation. *(Courtesy of Shannon & Wilson, Inc. Seattle, WA)*

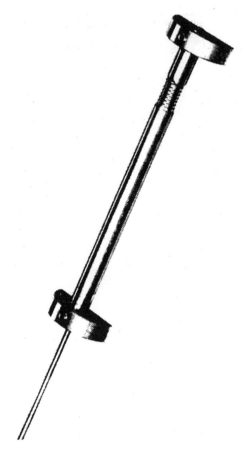

FIGURE 15.7 Strain gauges for strut monitoring. *(Courtesy of Slope Indicator Co. Mukilteo, WA)*

CHAPTER 16
OSHA REGULATIONS (STANDARDS—29CFR)

This chapter consists of the OSHA (Occupational Safety and Health Administration) Regulations 29 CFR (Code of Federal Regulations) 1926 subpart P—Excavation. These standards apply to construction, and in the absence of any overriding authority are the rules by which contractors must abide.

These rules are only applicable in the USA, and then only in the absence of individual state authority which may apply. A number of provinces in Canada have their own regulations, as do some states in the USA. WISHA (State of Washington, OROSHA (State of Oregon) and CalOSHA (State of California) are examples of regulatory codes which may take precedence over the Federal regulations. In general, state regulations in matters of construction safety will take precedence if they are more stringent than federal regulations.

The reader is cautioned to investigate all applicable regulations before utilizing the federal OSHA regulations which follow.

SCOPE, APPLICATION, AND DEFINITIONS APPLICABLE TO THIS SUBPART 1926.650

(a)

Scope and application. This subpart applies to all open excavations made in the earth's surface. Excavations are defined to include trenches.

(b)

Definitions applicable to this subpart.

Accepted engineering practices means those requirements which are compatible with standards of practice required by a registered professional engineer.

Aluminum Hydraulic Shoring means a pre-engineered shoring system comprised of aluminum hydraulic cylinders (crossbraces) used in conjunction with vertical rails (uprights) or horizontal rails (wales). Such system is designed specifically to support the sidewalls of an excavation and prevent cave-ins.

Bell-bottom pier hole means a type of shaft or footing excavation, the bottom of which is made larger than the cross section above to form a belled shape.

Benching (Benching system) means a method of protecting employees from cave-ins by excavating the sides of an excavation to form one or a series of horizontal levels or steps, usually with vertical or near-vertical surfaces between levels.

Cave-in means the separation of a mass of soil or rock material from the side of an excavation, or the loss of soil from under a trench shield or support system, and its sudden movement into the excavation, either by falling or sliding, in sufficient quantity so that it could entrap, bury, or other wise injure and immobilize a person.

Competent person means one who is capable of identifying existing and predictable hazards in the surroundings, or working conditions which are unsanitary, hazardous, or dangerous to employees, and who has authorization to take prompt corrective measures to eliminate them.

Cross braces mean the horizontal members of a shoring system installed perpendicular to the sides of the excavation, the ends of which bear against either uprights or wales.

Excavation means any man-made cut, cavity, trench, or depression in an earth surface, formed by earth removal.

Faces or *sides* means the vertical or inclined earth surfaces formed as a result of excavation work.

Failure means the breakage, displacement, or permanent deformation of a structural member or connection so as to reduce its structural integrity and its supportive capabilities.

Hazardous atmosphere means an atmosphere which by reason of being explosive, flammable, poisonous, corrosive, oxidizing, irritating, oxygen deficient, toxic, or otherwise harmful, may cause death, illness, or injury.

Kickout means the accidental release or failure of a cross brace.

Protective system means a method of protecting employees from cave-ins, from material that could fall or roll from an excavation face or into an excavation, or from the collapse of adjacent structures. Protective systems include support systems, sloping and benching systems, shield systems, and other systems that provide the necessary protection.

Ramp means an inclined walking or working surface that is used to gain access to one point from another, and is constructed from earth or from structural materials such as steel or wood.

Registered Professional Engineer means a person who is registered as a professional engineer in the state where the work is to be performed. However, a professional engineer, registered in any state is deemed to be a *registered professional engineer* within the meaning of this standard when approving designs for *manufactured protective systems* or *tabulated data* to be used in interstate commerce.

Sheeting means the members of a shoring system that retain the earth in position and in turn are supported by other members of the shoring system.

Shield (Shield system) means a structure that is able to withstand the forces imposed on it by a cave-in and thereby protect employees within the structure. Shields can be permanent structures or can be designed to be portable and moved along as work progresses. Additionally, shields can be either premanufactured or job-built in accordance with 1926.652(c)(3) or (c)(4). Shields used in trenches are usually referred to as *trench boxes* or *trench shields*.

Shoring (Shoring system) means a structure such as a metal hydraulic, mechanical or timber shoring system that supports the sides of an excavation and which is designed to prevent cave-ins.

Sides. See *Faces*.

Sloping (Sloping system) means a method of protecting employees from cave-ins by excavating to form sides of an excavation that are inclined away from the excavation so as to prevent cave-ins. The angle of incline required to prevent a cave-in varies with differences in such factors as the soil type, environmental conditions of exposure, and application of surcharge loads.

Stable rock means natural solid mineral material that can be excavated with vertical sides and will remain intact while exposed. Unstable rock is considered to be stable when the rock material on the side or sides of the excavation is secured against caving-in or movement by rock bolts or by another protective system that has been designed by a registered professional engineer.

Structural ramp means a ramp built of steel or wood, usually used for vehicle access. Ramps made of soil or rock are not considered structural ramps.

Support system means a structure such as underpinning, bracing, or shoring, which provides support to an adjacent structure, underground installation, or the sides of an excavation.

Tabulated data means tables and charts approved by a registered professional engineer and used to design and construct a protective system.

Trench (Trench excavation) means a narrow excavation (in relation to its length) made below the surface of the ground. In general, the depth is greater than the width, but the width of a trench (measured at the bottom) is not greater than 15 feet (4.6 m). If forms or other structures are installed or constructed in an excavation so as to reduce the dimension measured from the forms or structure to the side of the excavation to 15 feet (4.6 m) or less (measured at the bottom of the excavation), the excavation is also considered to be a trench.

Trench box. See *Shield.*

Trench shield. See *Shield.*

Uprights means the vertical members of a trench shoring system placed in contact with the earth and usually positioned so that individual members do not contact each other. Uprights placed so that individual members are closely spaced, in contact with or interconnected to each other, are often called *sheeting.*

Wales means horizontal members of a shoring system placed parallel to the excavation face whose sides bear against the vertical members of the shoring system or earth.

SPECIFIC EXCAVATION REQUIREMENTS—1926.651

(a)

Surface encumbrances. All surface encumbrances that are located so as to create a hazard to employees shall be removed or supported, as necessary, to safeguard employees.

(b)

Underground installations.

(b)(1)

The estimated location of utility installations, such as sewer, telephone, fuel, electric, water lines, or any other underground installations that reasonably may be expected to be encountered during excavation work, shall be determined prior to opening an excavation.

(b)(2)

Utility companies or owners shall be contacted within established or customary local response times, advised of the proposed work, and asked to establish the location of the utility underground installations prior to the start of actual excavation. When utility companies or owners cannot respond to a request to locate underground utility installations within 24 hours (unless a longer period is required by state or local law), or cannot establish the exact location of these installations, the employer may proceed, provided the employer does so with caution, and provided detection equipment or other acceptable means to locate utility installations are used.

(b)(3)

When excavation operations approach the estimated location of underground installations, the exact location of the installations shall be determined by safe and acceptable means.

(b)(4)

While the excavation is open, underground installations shall be protected, supported or removed as necessary to safeguard employees.

(c)

Access and egress.

(c)(1)

Structural ramps.

(c)(1)(i)

Structural ramps that are used solely by employees as a means of access or egress from excavations shall be designed by a competent person. Structural ramps used for access or egress of equipment shall be designed by a competent person qualified in structural design, and shall be constructed in accordance with the design.

(c)(1)(ii)

Ramps and runways constructed of two or more structural members shall have the structural members connected together to prevent displacement.

(c)(1)(iii)

Structural members used for ramps and runways shall be of uniform thickness.

(c)(1)(iv)

Cleats or other appropriate means used to connect runway structural members shall be attached to the bottom of the runway or shall be attached in a manner to prevent tripping.

(c)(1)(v)

Structural ramps used in lieu of steps shall be provided with cleats or other surface treatments of the top surface to prevent slipping.

(c)(2)

Means of egress from trench excavations. A stairway, ladder, ramp or other safe means of egress shall be located in trench excavations that are 4 feet (1.22 m) or more in depth so as to require no more than 25 feet (7.62 m) of lateral travel for employees.

(d)

Exposure to vehicular traffic. Employees exposed to public vehicular traffic shall be provided with, and shall wear, warning vests or other suitable garments marked with or made of reflectorized or high-visibility material.

(e)

Exposure to falling loads. No employee shall be permitted underneath loads handled by lifting or digging equipment. Employees shall be required to stand away from any vehicle being loaded or unloaded to avoid being struck by any spillage or falling materials. Operators may remain in the cabs of vehicles being loaded or unloaded when the vehicles are equipped, in accordance with f1926.601(b)(6), to provide adequate protection for the operator during loading and unloading operations.

(f)

Warning system for mobile equipment. When mobile equipment is operated adjacent to an excavation, or when such equipment is required to approach the edge of an excavation, and the operator does not have a clear and direct view of the edge of the excavation, a warning system shall be utilized such as barricades, hand or mechanical signals, or stop logs. If possible, the grade should be away from the excavation.

(g)

Hazardous atmospheres.

(g)(1)

Testing and controls. In addition to the requirements set forth in subparts D and E of this part (29 CFR 1926.50-1926.107) to prevent exposure to harmful levels of atmospheric contaminants and to assure acceptable atmospheric conditions, the following requirements shall apply:

(g)(1)(i)

Where oxygen deficiency (atmospheres containing less than 19.5 percent oxygen) or a hazardous atmosphere exists or could reasonably be expected to exist, such as in excavations in landfill areas or excavations in areas where hazardous substances are stored nearby, the atmospheres in the excavation shall be tested before employees enter excavations greater than 4 feet (1.22 m) fin depth.

(g)(1)(ii)

Adequate precautions shall be taken to prevent employee exposure to atmospheres containing less than 19.5 percent oxygen and other hazardous atmospheres. These precautions include providing proper respiratory protection or ventilation in accordance with subparts D and E of this part respectively.

(g)(1)(iii)

Adequate precaution shall be taken such as providing ventilation, to prevent employee exposure to an atmosphere containing a concentration of a flammable gas in excess of 20 percent of the flower flammable limit of the gas.

(g)(1)(iv)

When controls are used that are intended to reduce the level of atmospheric contaminants to acceptable levels, testing shall be conducted as often as necessary to ensure that the atmosphere remains safe.

(g)(2)

Emergency rescue equipment.

(g)(2)(i)

Emergency rescue equipment, such as breathing apparatus, a safety harness and line, or a basket stretcher, shall be readily available where hazardous atmospheric conditions exist or may reasonably be expected to develop during work in an excavation. This equipment shall be attended when in use.

(g)(2)(ii)

Employees entering bell-bottom pier holes, or other similar deep and confined footing excavations, shall wear a harness with a lifeline securely attached to it. The lifeline shall be separate from any line used to handle materials, and shall be individually attended at all times while the employee wearing the lifeline is in the excavation.

(h)

Protection from hazards associated with water accumulation.

(h)(1)

Employees shall not work in excavations in which there is accumulated water, or in excavations in which water is accumulating, unless adequate precautions have been taken to protect employees against the hazards posed by water accumulation. The precautions necessary to protect employees adequately vary with each situation, but could include special support or shield systems to protect from cave-ins, water removal to control the level of accumulating water, or use of a safety harness and lifeline.

(h)(2)

If water is controlled or prevented from accumulating by the use of water removal equipment, the water removal equipment and operations shall be monitored by a competent person to ensure proper operation.

(h)(3)

If excavation work interrupts the natural drainage of surface water (such as streams), diversion ditches, dikes, or other suitable means shall be used to prevent surface water from entering the excavation and to provide adequate drainage of the area adjacent to the excavation. Excavations subject to runoff from heavy rains will require an inspection by a competent person and compliance with paragraphs (h)(1) and (h)(2) of this section.

(i)

Stability of adjacent structures.

(i)(1)

Where the stability of adjoining buildings, walls, or other structures is endangered by excavation operations, support systems such as shoring, bracing, or underpinning shall be provided to ensure the stability of such structures for the protection of employees.

(i)(2)

Excavation below the level of the base or footing of any foundation or retaining wall that could be reasonably expected to pose a hazard to employees shall not be permitted except when:

(i)(2)(i)

A support system, such as underpinning, is provided to ensure the safety of employees and the stability of the structure; or

(i)(2)(ii)

The excavation is in stable rock; or

(i)(2)(iii)

A registered professional engineer has approved the determination that the structure is sufficiently removed from the excavation so as to be unaffected by the excavation activity; or

(i)(2)(iv)

A registered professional engineer has approved the determination that such excavation work will not pose a hazard to employees.

(i)(3)

Sidewalks, pavements and appurtenant structure shall not be undermined unless a support system or another method of protection is provided to protect employees from the possible collapse of such structures.

(j)

Protection of employees from loose rock or soil.

(j)(1)

Adequate protection shall be provided to protect employees from loose rock or soil that could pose a hazard by falling or rolling from an excavation face. Such protection shall consist of scaling to remove loose material; installation of protective barricades at intervals as necessary on the face to stop and contain falling material; or other means that provide equivalent protection.

(j)(2)

Employees shall be protected from excavated or other materials or equipment that could pose a hazard by falling or rolling into excavations. Protection shall be provided by placing and keeping such materials or equipment at least 2 feet (.61 m) from the edge of excavations, or by the use of retaining devices that are sufficient to prevent materials or equipment from falling or rolling into excavations, or by a combination of both if necessary.

(k)

Inspections.

(k)(1)

Daily inspections of excavations, the adjacent areas, and protective systems shall be made by a competent person for evidence of a situation that could result in possible cave-ins, indications of failure of protective systems, hazardous atmospheres, or other hazardous conditions. An inspection shall be conducted by the competent person prior to the start of work and as needed throughout the shift. Inspections shall also be made after every rainstorm or other hazard increasing occurrence. These inspections are only required when employee exposure can be reasonably anticipated.

(k)(2)

Where the competent person finds evidence of a situation that could result in a possible cave-in, indications of failure of protective systems, hazardous atmospheres, or other hazardous conditions, exposed employees shall be removed from the hazardous area until the necessary precautions have been taken to ensure their safety.

(l)

Fall protection.

(l)(1)

Walkways shall be provided where employees or equipment are required or permitted to cross over excavations. Guardrails which comply with 1926.502(b) shall be provided where walkways are 6 feet (1.8 m) or more above lower levels.

REQUIREMENTS FOR PROTECTIVE SYSTEMS—1926.652

a)

Protection of employees in excavations.

(a)(1)

Each employee in an excavation shall be protected from cave-ins by an adequate protective system designed in accordance with paragraph (b) or (c) of this section except when:

(a)(1)(i)

Excavations are made entirely in stable rock; or

(a)(1)(ii)

Excavations are less than 5 feet (1.52 m) in depth and examination of the ground by a competent person provides no indication of a potential cave-in.

(a)(2)

Protective systems shall have the capacity to resist without failure all loads that are intended or could reasonably be expected to be applied or transmitted to the system.

(b)

Design of sloping and benching systems. The slopes and configurations of sloping and benching systems shall be selected and constructed by the employer or his designee and shall be in accordance with the requirements of paragraph (b)(1); or, in the alternative, paragraph (b)(2); or, in the alternative, paragraph (b)(3); or, in the alternative, paragraph (b)(4), as follows:

(b)(1)

Option (1)—Allowable configurations and slopes.

(b)(1)(i)

Excavations shall be sloped at an angle not steeper than one and one-half horizontal to one vertical (34 degrees measured from the horizontal), unless the employer uses one of the other options listed below.

(b)(1)(ii)

Slopes specified in paragraph (b)(1)(i) of this section, shall be excavated to form configurations that are in accordance with the slopes shown for Type C soil in Appendix B to this subpart.

(b)(2)

Option (2)—Determination of slopes and configurations using Appendices A and B. Maximum allowable slopes, and allowable configurations for sloping and benching systems, shall be determined in accordance with the conditions and requirements set forth in appendices A and B to this subpart.

(b)(3)

Option (3)—Designs using other tabulated data.

(b)(3)(i)

Designs of sloping or benching systems shall be selected from and in accordance with tabulated data, such as tables and charts.

(b)(3)(ii)

The tabulated data shall be in written form and shall include all of the following:

(b)(3)(ii)(A)

Identification of the parameters that affect the selection of a sloping or benching system drawn from such data;

(b)(3)(ii)(B)

Identification of the limits of use of the data, to include the magnitude and configuration of slopes determined to be safe;

(b)(3)(ii)(C)

Explanatory information as may be necessary to aid the user in making a correct selection of a protective system from the data.

(b)(3)(iii)

At least one copy of the tabulated data which identifies the registered professional engineer who approved the data, shall be maintained at the jobsite during construction of the protective system. After that time the data may be stored off the jobsite, but a copy of the data shall be made available to the Secretary upon request.

(b)(4)

Option (4)—Design by a registered professional engineer.

(b)(4)(i)

Sloping and benching systems not utilizing Option (1) or Option (2) or Option (3) under paragraph (b) of this section shall be approved by a registered professional engineer.

(b)(4)(ii)

Designs shall be in written form and shall include at least the following:

(b)(4)(ii)(A)

The magnitude of the slopes that were determined to be safe for the particular project;

(b)(4)(ii)(B)

The configurations that were determined to be safe for the particular project;

(b)(4)(ii)(C)

The identity of the registered professional engineer approving the design.

(b)(4)(iii)

At least one copy of the design shall be maintained at the jobsite while the slope is being constructed. After that time the design need not be at the jobsite, but a copy shall be made available to the Secretary upon request.

(c)

Design of support systems, shield systems, and other protective systems. Designs of support systems, shield systems, and other protective systems shall be selected and constructed by the employer or his designee and shall be in accordance with the requirements of paragraph (c)(1); or, in the alternative, paragraph (c)(2); or, in the alternative, paragraph (c)(3); or, i the alternative, paragraph (c)(4) as follows:

(c)(1)

Option (1)—Designs using appendices A, C and D. Designs for timber shoring in trenches shall be determined in accordance with the conditions and requirements set forth in appendices A and C to this subpart. Designs for aluminum hydraulic shoring shall be in accordance with paragraph (c)(2) of this section, but if manufacturer's tabulated data cannot be utilized, designs shall be in accordance with appendix D.

(c)(2)

Option (2)—Designs Using Manufacturer's Tabulated Data.

(c)(2)(i)

Design of support systems, shield systems, or other protective systems that are drawn from manufacturer's tabulated data shall be in accordance with all specifications, recommendations, and limitations issued or made by the manufacturer.

(c)(2)(ii)

Deviation from the specifications, recommendations, and limitations issued or made by the manufacturer shall only be allowed after the manufacturer issues specific written approval.

(c)(2)(iii)

Manufacturer's specifications, recommendations, and limitations, and manufacturer's approval to deviate from the specifications, recommendations, and limitations shall be in written form at the jobsite during construction of the protective system. After that time this data may be stored off the jobsite, but a copy shall be made available to the Secretary upon request.

(c)(3)

Option (3)—Designs using other tabulated data.

(c)(3)(i)

Designs of support systems, shield systems, or other protective systems shall be selected from and be in accordance with tabulated data, such as tables and charts.

(c)(3)(ii)

The tabulated data shall be in written form and include all of the following:

(c)(3)(ii)(A)

Identification of the parameters that affect the selection of a protective system drawn from such data;

(c)(3)(ii)(B)

Identification of the limits of use of the data;

(c)(3)(ii)(C)

Explanatory information as may be necessary to aid the user in making a correct selection of a protective system from the data.

(c)(3)(iii)

At least one copy of the tabulated data, which identifies the registered professional engineer who approved the data, shall be maintained at the jobsite during construction of the protective system. After that time the data may be stored off the jobsite, but a copy of the data shall be made available to the Secretary upon request.

(c)(4)

Option (4)—Design by a registered professional engineer.

(c)(4)(i)

Support systems, shield systems, and other protective systems not utilizing Option 1, Option 2, or Option 3, above, shall be approved by a registered professional engineer.

(c)(4)(ii)

Designs shall be in written form and shall include the following:

(c)(4)(ii)(A)

A plan indicating the sizes, types, and configurations of the materials to be used in the protective system; and

(c)(4)(ii)(B)

The identify of the registered professional engineer approving the design.

(c)(4)(iii)

At least one copy of the design shall be maintained at the jobsite during construction of the protective system. After that time, the design may be stored off the jobsite, but a copy of the design shall be made available to the Secretary upon request.

(d)

Materials and equipment.

(d)(1)

Materials and equipment used for protective systems shall be free from damage or defects that might impair their proper function.

(d)(2)

Manufactured materials and equipment used for protective systems shall be used and maintained in a manner that is consistent with the recommendations of the manufacturer, and in a manner that will prevent employee exposure to hazards.

(d)(3)

When material or equipment that is used for protective systems is damaged, a competent person shall examine the material or equipment and evaluate its suitability for continued use. If the competent person cannot assure the material or equipment is able to support the intended loads or is otherwise suitable for safe use, then such material or equipment shall be removed from service, and shall be evaluated and approved by a registered professional engineer before being returned to service.

(e)

Installation and removal of support.

(e)(1)

General.

(e)(1)(i)

Members of support systems shall be securely connected together to prevent sliding, falling, kickouts, or other predictable failure.

(e)(1)(ii)

Support systems shall be installed and removed in a manner that protects employees from cave-ins, structural collapses, or from being struck by members of the support system.

(e)(1)(iii)

Individual members of support systems shall not be subjected to loads exceeding those which those members were designed to withstand.

(e)(1)(iv)

Before temporary removal of individual members begins, additional precautions shall be taken to ensure the safety of employees, such as installing other structural members to carry the loads imposed on the support system.

(e)(1)(v)

Removal shall begin at, and progress from, the bottom of the excavation. Members shall be released slowly so as to note any indication of possible failure of the remaining members of the structure or possible cave-in of the sides of the excavation.

(e)(1)(vi)

Backfilling shall progress together with the removal of support systems from excavations.

(e)(2)

Additional requirements for support systems for trench excavations.

(e)(2)(i)

Excavation of material to a level no greater than 2 feet (.61 m) below the bottom of the members of a support system shall be permitted, but only if the system is designed to resist the forces calculated for the full depth of the trench, and there are no indications while the trench is open of a possible loss of soil from behind or below the bottom of the support system.

(e)(2)(ii)

Installation of a support system shall be closely coordinated with the excavation of trenches.

(f)

Sloping and benching systems. Employees shall not be permitted to work on the faces of sloped or benched excavations at levels above other employees except when employees at the lower levels are adequately protected from the hazard of falling, rolling, or sliding material or equipment.

(g)

Shield systems

(g)(1)

General.

(g)(1)(i)

Shield systems shall not be subjected to loads exceeding those which the system was designed to withstand.

(g)(1)(ii)

Shields shall be installed in a manner to restrict lateral or other hazardous movement of the shield in the event of the application of sudden lateral loads.

(g)(1)(iii)

Employees shall be protected from the hazard of cave-ins when entering or exiting the areas protected by shields.

(g)(1)(iv)

Employees shall not be allowed in shields when shields are being installed, removed, or moved vertically.

(g)(2)

Additional requirement for shield systems used in trench excavations. Excavations of earth material to a level not greater than 2 feet (.61 m) below the bottom of a shield shall be permitted, but only if the shield is designed to resist the forces calculated for the full depth of the trench, and there are no indications while the trench is open of a possible loss of soil from behind or below the bottom of the shield.

AUTHORITY FOR 1926 SUBPART P—1926 SUBPART P

Authority: Sec. 107, Contract Worker Hours and Safety Standards Act (Construction Safety Act)

(40 U.S.C. 333); Secs. 4, 6, 8, Occupational Safety and Health Act of 1970 (29 U.S.C. 653, 655,

657); Secretary of Labor's Order No. 12-71 (36 FR 8754), 8-76 (41 FR 25059), 9-83 (48 FR

35736), or 1-90 (55 FR 9033), as applicable.

Section 1926.651 also issued under 29 CFR Part 1911.

Source: 54 FR 45959, Oct. 31, 1989, unless otherwise noted.

SOIL CLASSIFICATION—1926 SUBPART P APP A

(a)

Scope and application

(a)(1)

Scope. This appendix describes a method of classifying soil and rock deposits based on site and environmental conditions, and on the structure and composition of the earth deposits. The appendix contains definitions, sets forth requirements, and describes acceptable visual and manual tests for use in classifying soils.

(a)(2)

Application. This appendix applies when a sloping or benching system is designed in accordance with the requirements set forth in 1926.652(b)(2) as a method of protection for employees from cave-ins. This appendix also applies when timber shoring for excavations is designed as a method of protection from cave-ins in accordance with appendix C to subpart P of part 1926, and when aluminum hydraulic shoring is designed in accordance with appendix D. This Appendix also applies if other protective systems are designed and selected for use from data prepared in accordance with the requirements set forth in 1926.652(c), and the use of the data is predicated on the use of the soil classification system set forth in this appendix.

(b)

Definitions. The definitions and examples given below are based on, in whole or in part, the following; American Society for Testing Materials (ASTM) Standards D653-85 and D2488; The Unified Soils Classification System; The U.S. Department of Agriculture (USDA) Textural Classification Scheme; and The National Bureau of Standards Report BSS-121.

Cemented soil means a soil in which the particles are held together by a chemical agent, such as calcium carbonate, such that a hand-size sample cannot be crushed into powder or individual soil particles by finger pressure.

Cohesive soil means clay (fine grained soil), or soil with a high clay content, which has cohesive strength. Cohesive soil does not crumble, can be excavated with vertical sidesloper, and is plastic when moist. Cohesive soil is hard to break up when dry, and exhibits significant cohesion when submerged. Cohesive soils include clayey silt, sandy clay, silty clay, clay and organic clay.

Dry soil means soil that does not exhibit visible signs of moisture content.

Fissured means a soil material that has a tendency to break along definite planes of fracture with little resistance, or a material that exhibits open cracks, such as tension cracks, in an exposed surface.

Granular soil means gravel, sand, or silt (coarse grained soil) with little or no clay content. Granular soil has no cohesive strength. Some moist granular soils exhibit apparent cohesion. Granular soil cannot be molded when moist and crumbles easily when dry.

Layered system means two or more distinctly different soil or rock types arranged in layers.

Micaceous seams or weakened planes in rock or shale are considered layered.

Moist soil means a condition in which a soil looks and feels damp. Moist cohesive soil can easily be shaped into a ball and rolled into small diameter threads before crumbling. Moist granular soil that contains some cohesive material will exhibit signs of cohesion between particles.

Plastic means a property of a soil which allows the soil to be deformed or molded without cracking, or appreciable volume change.

Saturated soil means a soil in which the voids are filled with water. Saturation does not require flow. Saturation, or near saturation, is necessary for the proper use of instruments such as a pocket penetrometer or sheer vane.

Soil classification system means, for the purpose of this subpart, a method of categorizing soil and rock deposits in a hierarchy of Stable Rock, Type A, Type B, and Type C, in decreasing order of stability. The categories are determined based on an analysis of the properties and performance characteristics of the deposits and the characteristics of the deposits and the environmental conditions of exposure.

Stable rock means natural solid mineral matter that can be excavated with vertical sides and remain intact while exposed.

Submerged soil means soil which is underwater or is free seeping. Type A means cohesive soils with an unconfined, compressive strength of 1.5 ton per square foot (tsf) (144 kPa) or greater. Examples of cohesive soils are: clay, silty clay, sandy clay, clay loam and, in some cases, silty clay loam and sandy clay loam.

Cemented soils such as caliche and hardpan are also considered Type A. However, no soil is Type A if:

(i) The soil is fissured; or

(ii) The soil is subject to vibration from heavy traffic, pile driving, or similar effects; or

(iii) The soil has been previously disturbed; or

(iv) The soil is part of a sloped, layered system where the layers dip into the excavation on a slope of four horizontal to one vertical (4H:1V) or greater; or

(v) The material is subject to other factors that would require it to be classified as a less stable material.

Type B means:

(i) Cohesive soil with an unconfined compressive strength greater than 0.5 tsf (48 kPa) but less than 1.5 tsf (144 kPa); or

(ii) Granular cohesionless soils including: angular gravel (similar to crushed rock), silt, silt loam, sandy loam and, in some cases, silty clay loam and sandy clay loam.

(iii) Previously disturbed soils except those which would otherwise be classed as Type C soil.

(iv) Soil that meets the unconfined compressive strength or cementation requirements for Type A, but is fissured or subject to vibration; or

(v) Dry rock that is not stable; or

(vi) Material that is part of a sloped, layered system where the layers dip into the excavation on a slope less steep than four horizontal to one vertical (4H:1V), but only if the material would otherwise be classified as Type B.

Type C means:

(i) Cohesive soil with an unconfined compressive strength of 0.5 tsf (48 kPa) or less; or

(ii) Granular soils including gravel, sand, and loamy sand; or

(iii) Submerged soil or soil from which water is freely seeping; or

(iv) Submerged rock that is not stable, or

(v) Material in a sloped, layered system where the layers dip into the excavation or a slope of four horizontal to one vertical (4H:1V) or steeper.

Unconfined compressive strength means the load per unit area at which a soil will fail in compression. It can be determined by laboratory testing, or estimated in the field using a pocket penetrometer, by thumb penetration tests, and other methods.

Wet soil means soil that contains significantly more moisture than moist soil, but in such a range of values that cohesive material will slump or begin to flow when vibrated. Granular material that would exhibit cohesive properties when moist will lose those cohesive properties when wet.

(c)

Requirements

(c)(1)

Classification of soil and rock deposits. Each soil and rock deposit shall be classified by a competent person as Stable Rock, Type A, Type B, or Type C in accordance with the definitions set forth in paragraph (b) of this appendix.

(c)(2)

Basis of classification. The classification of the deposits shall be made based on the results of at least one visual and at least one manual analysis. Such analyses shall be conducted by a competent person using tests described in paragraph (d) below, or in other recognized methods of soil classification and testing such as those adopted by the American Society for Testing Materials, or the U.S. Department of Agriculture textural classification system.

(c)(3)

Visual and manual analyses. The visual and manual analyses, such as those noted as being acceptable in paragraph (d) of this appendix, shall be designed and conducted to provide sufficient quantitative and qualitative information as may be necessary to identify properly the properties, factors, and conditions affecting the classification of the deposits.

(c)(4)

Layered systems. In a layered system, the system shall be classified in accordance with its weakest layer. However, each layer may be classified individually where a more stable layer lies under a less stable layer.

(c)(5)

Reclassification. If, after classifying a deposit, the properties, factors, or conditions affecting its classification change in any way, the changes shall be evaluated by a competent person. The deposit shall be reclassified as necessary to reflect the changed circumstances.

(d)

Acceptable visual and manual tests.

(d)(1)

Visual tests. Visual analysis is conducted to determine qualitative information regarding the excavation site in general, the soil adjacent to the excavation, the soil forming the sides of the open excavation, and the soil taken as samples from excavated material.

(d)(1)(i)

Observe samples of soil that are excavated and soil in the sides of the excavation. Estimate the range of particle sizes and the relative amounts of the particle sizes. Soil that is primarily composed of fine-grained material material is cohesive material. Soil composed primarily of coarse-grained sand or gravel is granular material.

(d)(1)(ii)

Observe soil as it is excavated. Soil that remains in clumps when excavated is cohesive. Soil that breaks up easily and does not stay in clumps is granular.

(d)(1)(iii)

Observe the side of the opened excavation and the surface area adjacent to the excavation. Crack-like openings such as tension cracks could indicate fissured material. If chunks of soil spall off a vertical side, the soil could be fissured. Small spalls are evidence of moving ground and are indications of potentially hazardous situations.

(d)(1)(iv)

Observe the area adjacent to the excavation and the excavation itself for evidence of existing utility and other underground structures, and to identify previously disturbed soil.

(d)(1)(v)

Observe the opened side of the excavation to identify layered systems. Examine layered systems to identify if the layers slope toward the excavation. Estimate the degree of slope of the layers.

(d)(1)(vi)

Observe the area adjacent to the excavation and the sides of the opened excavation for evidence of surface water, water seeping from the sides of the excavation, or the location of the level of the water table.

(d)(1)(vii)

Observe the area adjacent to the excavation and the area within the excavation for sources of vibration that may affect the stability of the excavation face.

(d)(2)

Manual tests. Manual analysis of soil samples is conducted to determine quantitative as well as qualitative properties of soil and to provide more information in order to classify soil properly.

(d)(2)(i)

Plasticity. Mold a moist or wet sample of soil into a ball and attempt to roll it into threads as thin as ⅛ inch in diameter. Cohesive material can be successfully rolled into threads without crumbling. For example, if at least a two inch (50 mm) length of ⅛ inch thread can be held on one end without tearing, the soil is cohesive.

(d)(2)(ii)

Dry strength. If the soil is dry and crumbles on its own or with moderate pressure into individual grains or fine powder, it is granular (any combination of gravel, sand, or silt). If the soil is dry and falls into clumps which break up into smaller clumps, but the smaller clumps can only be broken up with difficulty, it may be clay in any combination with gravel, sand or silt. If the dry soil breaks into clumps which do not break up into small clumps and which can only be broken with difficulty, and there is no visual indication the soil is fissured, the soil may be considered unfissured.

(d)(2)(iii)

Thumb penetration. The thumb penetration test can be used to estimate the unconfined compressive strength of cohesive soils. (This test is based on the thumb penetration test described in American Society for Testing and Materials (ASTM) Standard designation D2488.)

"Standard Recommended Practice for Description of Soils (Visual—Manual Procedure). Type A soils with an unconfined compressive strength of 1.5 tsf can be readily indented by the thumb; however, they can be penetrated by the thumb only with very great effort. Type C soils with an unconfined compressive strength of 0.5 tsf can be easily penetrated several inches by the thumb, and can be molded by light finger pressure. This test should be conducted on an undisturbed soil sample, such as a large clump of spoil, as soon as practicable after excavation to keep to a minimum the effects of exposure to drying influences. If the excavation is later exposed to wetting influences (rain, flooding), the classification of the soil must be changed accordingly.

(d)(2)(iv)

Other strength tests. Estimates of unconfined compressive strength of soils can also be obtained by use of a pocket penetrometer or by using a hand-operated shearvane.

(d)(2)(v)

Drying test. The basic purpose of the drying test is to differentiate between cohesive material with fissures, unfissured cohesive material, and granular material. The procedure for the drying test involves drying a sample of soil that is approximately one inch thick (2.54 cm) and six inches (15.24 cm) in diameter until it is thoroughly dry:

(d)(2)(v)(A)

If the sample develops cracks as it dries, significant fissures are indicated.

(d)(2)(v)(B)

Samples that dry without cracking are to be broken by hand. If considerable force is necessary to break a sample, the soil has significant cohesive material content. The soil can be classified as an unfissured cohesive material and the unconfined compressive strength should be determined.

(d)(2)(v)(C)

If a sample breaks easily by hand, it is either a fissured cohesive material or a granular material. To distinguish between the two, pulverize the dried clumps of the sample by hand or by stepping on them. If the clumps do not pulverize easily, the material is cohesive with fissures. If they pulverize easily into very small fragments, the material is granular.

SLOPING AND BENCHING—1926 SUBPART P APP B

(a)

Scope and application. This appendix contains specifications for sloping and benching when used as methods of protecting employees working in excavations from cave-ins. The requirements of this appendix apply when the design of sloping and benching protective systems is to be performed in accordance with the requirements set forth in 1926.652(b)(2).

(b)

Definitions.

Actual slope means the slope to which an excavation face is excavated.

Distress means that the soil is in a condition where a cave-in is imminent or is likely to occur. Distress is evidenced by such phenomena as the development of fissures in the face of or adjacent to an open excavation; the subsidence of the edge of an excavation; the slumping of material from the face or the bulging or heaving of material from the bottom of an excavation; the spalling of material from the face of an excavation; and ravelling, i.e., small amounts of material such as pebbles or little clumps of material suddenly separating from the face of an excavation and trickling or rolling down into the excavation.

Maximum allowable slope means the steepest incline of an excavation face that is acceptable for the most favorable site conditions as protection against cave-ins, and is expressed as the ratio of horizontal distance to vertical rise (H:V).

Short term exposure means a period of time less than or equal to 24 hours that an excavation is open.

(c)

Requirements.

(c)(1)

Soil classification. Soil and rock deposits shall be classified in accordance with appendix A to subpart P of part 1926.

(c)(2)

Maximum allowable slope. The maximum allowable slope for a soil or rock deposit shall be determined from Table B-1 of this appendix.

(c)(3)

Actual slope.

(c)(3)(i)

The actual slope shall not be steeper than the maximum allowable slope.

(c)(3)(ii)

The actual slope shall be less steep than the maximum allowable slope, when there are signs of distress. If that situation occurs, the slope shall be cut back to an actual slope which is at least ½ horizontal to one vertical (1/2H:1V) less steep than the maximum allowable slope.

(c)(3)(iii)

When surcharge loads from stored material or equipment, operating equipment, or traffic are present, a competent person shall determine the degree to which the actual slope must be reduced below the maximum allowable slope, and shall assure that such reduction is achieved. Surcharge loads from adjacent structures shall be evaluated in accordance with 1926.651(i).

(c)(4)

Configurations. Configurations of sloping and benching systems shall be in accordance with Figure B-1.

B.1.1 Excavations Made in Type A Soil

1. All simple slope excavation 20 feet or less in depth shall have a maximum allowable slope of 3/4:1.

TABLE B-1
MAXIMUM ALLOWABLE SLOPES

SOIL OR ROCK TYPE	MAXIMUM ALLOWABLE SLOPES (H:V)(1) FOR EXCAVATIONS LESS THAN 20 FEET DEEP(3)
STABLE ROCK	VERTICAL (90 Deg.)
TYPE A (2)	3/4:1 (53 Deg.)
TYPE B	1:1 (45 Deg.)
TYPE C	1 1/2:1 (34 Deg.)

Footnote(1) Numbers shown in parentheses next to maximum allowable slopes are angles expressed in degrees from the horizontal. Angles have been rounded off.

Footnote(2) A short-term maximum allowable slope of 1/2H:1V (63 degrees) is allowed in excavations in Type A soil that are 12 feed (3.67 m) or less in depth. Short-term maximum allowable slopes for excavations greater than 12 feet (3.67 m) in depth shall be 3/4H:1V (53 degrees).

Footnote(3) Sloping or benching for excavations greater than 20 feet deep shall be designed by a registered professional engineer.

FIGURE B-1 Slope configurations.

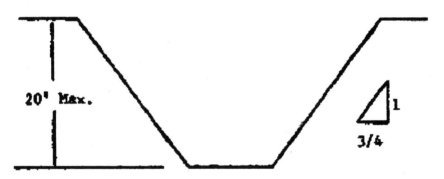

FIGURE B-1.1a Simple slope—general.

Exception: Simple slope excavations which are open 24 hours or less (short term) and which are 12 feet or less in depth shall have a maximum allowable slope of 1/2:1.

2. All benched excavations 20 feet or less in depth shall have a maximum allowable slope of ¾ to 1 and maximum bench dimensions as follows:

3. All excavations 8 feet or less in depth which have unsupported vertically sided lower portions shall have a maximum vertical side of 3½ feet.

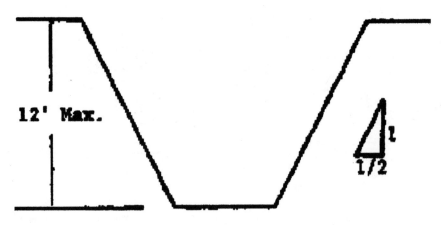

FIGURE B-1.1b Simple slope—short term.

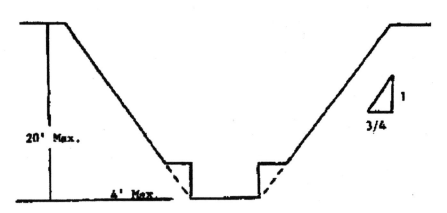

FIGURE B-1.1c Simple bench.

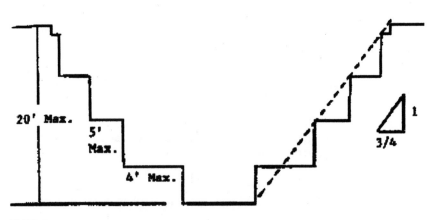

FIGURE B-1.1d Multiple bench.

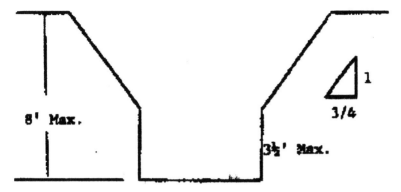

FIGURE B-1.1e Unsupported vertically sided lower portion—maximum 8 feet in depth.

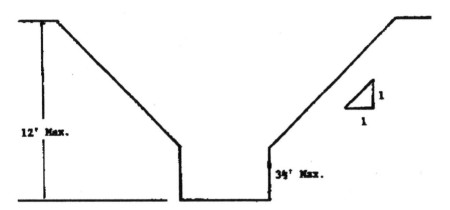

FIGURE B-1.1f Unsupported vertically sided lower portion—maximum 12 feet in depth.

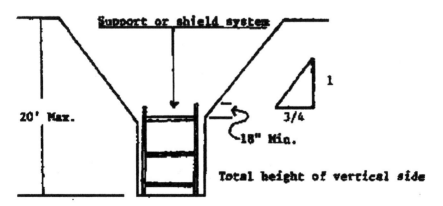

FIGURE B-1.1g Supported or shielded vertically sided lower portion.

All excavations more than 8 feet but not more than 12 feet in depth with unsupported vertically sided lower portions shall have a maximum allowable slope of 1:1 and a maximum vertical side of 3½ feet.

All excavations 20 feet or less in depth which have vertically sided lower portions that are supported or shielded shall have a maximum allowable slope of 3/4:1. The support or shield system must extend at least 18 inches above the top of the vertical side.

4. All other simple slope, compound slope, and vertically sided lower portion excavations shall be in accordance with the other options permitted under 1926.652(b).

B.1.2 Excavations Made in Type B Soil

1. All simple slope excavations 20 feet or less in depth shall have a maximum allowable slope of 1:1.
2. All benched excavations 20 feet or less in depth shall have a maximum allowable slope of 1:1 and maximum bench dimensions as follows:
3. All excavations 20 feet or less in depth which have vertically sided lower portions shall be shielded or supported to a height at least 18 inches above the top of the vertical side. All such excavations shall have a maximum allowable slope of 1:1.
4. All other sloped excavations shall be in accordance with the other options permitted in 1926.652(b).

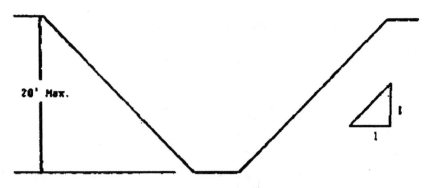

FIGURE B-1.2a Simple slope.

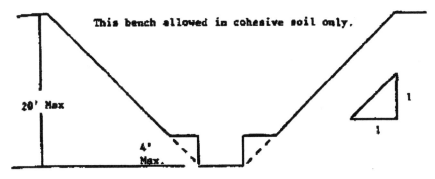

FIGURE B-1.2b Single bench.

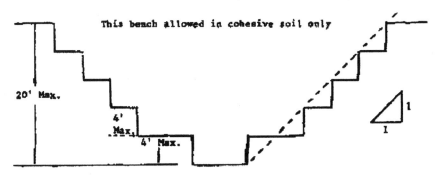

FIGURE B-1.2c Multiple bench.

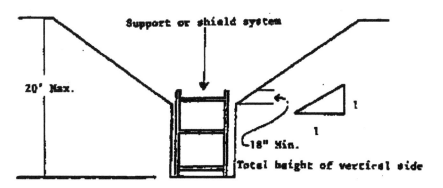

FIGURE B-1.2d Vertically sided lower portion.

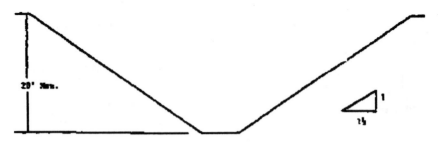

FIGURE B-1.3a Simple slope.

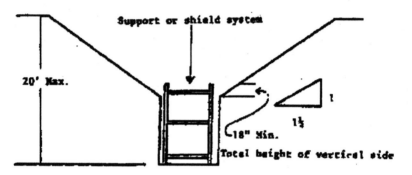

FIGURE B-1.3b Vertical sided lower portion.

B.1.3 Excavations Made in Type C Soil

1. All simple slope excavations 20 feet or less in depth shall have a maximum allowable slope of 1 1/2:1.
2. All excavations 20 feet or less in depth which have vertically sided lower portions shall be shielded or supported to a height at least 18 inches above the top of the vertical side. All such excavations shall have a maximum allowable slope of 1 1/2:1.
3. All other sloped excavations shall be in accordance with the other options permitted in 1926.652(b).

B.1.4 Excavations Made in Layered Soils

1. All excavations 20 feet or less in depth made in layered soils shall have a maximum allowable slope for each layer as set forth as follows.
2. All other sloped excavations shall be in accordance with the other options permitted in 1926.652(b).

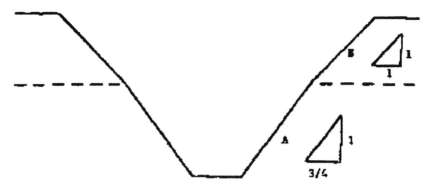

FIGURE B-1.4a B over A.

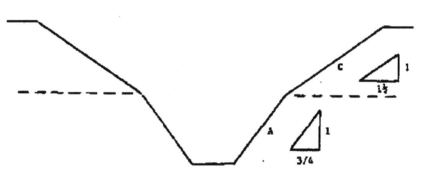

FIGURE B-1.4b C over A.

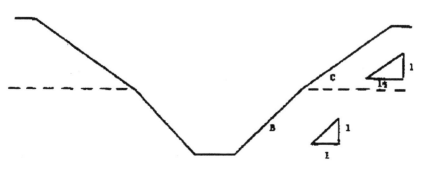

FIGURE B-1.4c C over B.

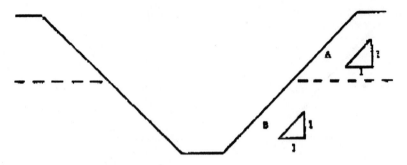

FIGURE B-1.4d A over B.

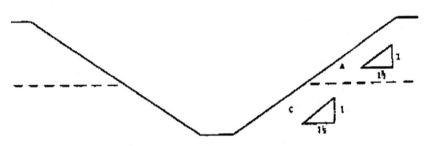

FIGURE B-1.4e A over C.

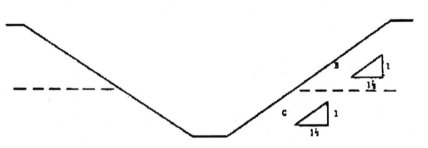

FIGURE B-1.4f B over C.

TIMBER SHORING FOR TRENCHES—1926 SUBPART P APP C

(a)

Scope. This appendix contains information that can be used when timber shoring is provided as a method of protection from cave-ins in trenches that do not exceed 20 feet (6.1 m) in depth. This appendix must be used when design of timber shoring protective systems is to be performed in accordance with 1926.652(c)(1). Other timber shoring configurations; other systems of support such as hydraulic and pneumatic systems; and other protective systems such as sloping, benching, shielding, and freezing systems must be designed in accordance with the requirements set forth in 1926.652(b) and 1926.652(c).

(b)

Soil Classification. In order to use the data presented in this appendix, the soil type or types in which the excavation is made must first be determined using the soil classification method set forth in appendix A of subpart P of this part.

(c)

Presentation of Information. Information is presented in several forms as follows:

(c)(1)

Information is presented in tabular form in Tables C-1.1, C-1.2 and C-1.3, and Tables C-2.1, C-2.2 and C-2.3 following paragraph (g) of the appendix. Each table presents the minimum sizes of timber members to use in a shoring system, and each table contains data only for the particular soil type in which the excavation or portion of the excavation is made. The data are arranged to allow the user the flexibility to select from among several acceptable configurations of members based on varying the horizontal spacing of the crossbraces. Stable rock is exempt from shoring requirements and therefore, no data are presented for this condition.

(c)(2)

Information concerning the basis of the tabular data and the limitations of the data is presented in paragraph (d) of this appendix, and on the tables themselves.

(c)(3)

Information explaining the use of the tabular data is presented in paragraph (e) of this appendix.

(c)(4)

Information illustrating the use of the tabular data is presented in paragraph (f) of this appendix.

(c)(5)

Miscellaneous notations regarding Tables C-1.1 through C-1.3 and Tables C-2.1 through C-2.3 are presented in paragraph (g) of this Appendix.

(d)

Basis and limitations of the data.

(d)(1)

Dimensions of timber members.

(d)(1)(i)

The sizes of the timber members listed in Tables C-1.1 through C-1.3 are taken from the National Bureau of Standards (NBS) report, "Recommended Technical Provisions for Construction Practice in Shoring and Sloping of Trenches and Excavations." In addition, where NBS did not recommend specific sizes of members, member sizes are based on an analysis of the sizes required for use by existing codes and on empirical practice.

(d)(1)(ii)

The required dimensions of the members listed in Tables C-1.1 through C-1.3 refer to actual dimensions and not nominal dimensions of the timber. Employers wanting to use nominal size shoring are directed to Tables C-2.1 through C-2.3, or have this choice under 1926.652(c)(3), and are referred to The Corps of engineers, The Bureau of Reclamation or data from other acceptable sources.

(d)(2)

Limitation of application.

(d)(2)(i)

it is not intended that the timber shoring specification apply to every situation that may be experienced in the field. These data were developed to apply to the situations that are most commonly experienced in current trenching practice. Shoring systems for use in situations that are not covered by the data in this appendix must be designed as specified in 1926.652(c).

(d)(2)(ii)

When any of the following conditions are present, the members specified in the tables are not considered adequate. Either an alternate timber shoring system must be designed or another type of protective system designed in accordance with 1926.652.

(d)(2)(ii)(A)

When loads imposed by structures or by stored material adjacent to the trench weigh in excess of the load imposed by a two-foot soil surcharge. The term "adjacent" as used here means the area within a horizontal distance from the edge of the trench equal to the depth of the trench.

(d)(2)(ii)(B)

When vertical loads imposed on cross braces exceed a 240-pound gravity load distributed on a one-foot section of the center of the crossbrace.

(d)(2)(ii)(C)

When surcharge loads are present from equipment weighing in excess of 20,000 pounds.

(d)(2)(ii)(D)

When only the lower portion of a trench is shored and the remaining portion of the trench is sloped or benched unless: The sloped portion is sloped at an angle less steep than three horizontal to one vertical; or the members are selected from the tables for use at a depth which is determined from the top of the overall trench, and not from the toe of the sloped portion.

(e)

Use of Tables. The members of the shoring system that are to be selected using this information are the cross braces, the uprights, and the wales, where wales are required. Minimum sizes of members are specified for use in different types of soil. There are six tables of information, two for each soil type. The soil type must first be determined in accordance with the soil classification system described in appendix A to subpart P of part 1926. Using the appropriate table, the selection of the size and spacing of the members is then made. The selection is based on the depth and width of the trench where the members are to be installed and, in most instances, the selection is also based on the horizontal spacing of the crossbraces. Instances where a choice of horizontal spacing of crossbracing is available, the horizontal spacing of the crossbraces must be chosen by the user before the size of any member can be determined. When the soil type, the width and depth of the trench, and the horizontal spacing of the crossbraces are known, the size and vertical spacing of the crossbraces are known, the size and vertical spacing of the crossbraces, the size and vertical spacing of the wales, and the size and horizontal spacing of the uprights can be read from the appropriate table.

(f)

Examples to Illustrate the Use of Tables C-1.1 through C-1.3.

(f)(1) Example 1

A trench dug in Type A soil is 13 feet deep and five feet wide. From Table C-1.1, for acceptable arrangements of timber can be used.

Arrangement #1
Space 4 x 4 crossbraces at six feet horizontally and four feet vertically.
Wales are not required.
Space 3 x 8 uprights at six feet horizontally. This arrangement is commonly called *skip shoring*.

Arrangement #2
Space 4 x 6 crossbraces at eight feet horizontally and four feet vertically.
Space 8 x 8 wales at four feet vertically.
Space 2 x 6 uprights at four feet horizontally.

Arrangement #3
Space 6 x 6 crossbraces at 10 feet horizontally and four feet vertically.
Space 8 x 10 wales at four feet vertically.
Space 2 x 6 uprights at five feet horizontally.

Arrangement #4
Space 6 x 6 crossbraces at 12 feet horizontally and four feet vertically.

Space 10 x 10 wales at four feet vertically.
Space 3 x 8 uprights at six feet horizontally.

(f)(2) Example 2

A trench dug in Type B soil is 13 feet deep and five feet wide. From Table C-1.2 three acceptable arrangements of members are listed.

Arrangement #1
Space 6 x 6 crossbraces at six feet horizontally and five feet vertically.
Space 8 x 8 wales at five feet vertically.
Space 2 x 6 uprights at two feet horizontally.

Arrangement #2
Space 6 x 8 crossbraces at eight feet horizontally and five feet vertically.
Space 10 x 10 wales at five feet vertically.
Space 2 x 6 uprights at two feet horizontally.

Arrangement #3
Space 8 x 8 crossbraces at 10 feet horizontally and five feet vertically.
Space 10 x 12 wales at five feet vertically.
Space 2 x 6 uprights at two feet vertically.

(f)(3) Example 3

A trench dug in Type C soil is 13 feet deep and five feet wide. From Table C-1.3 two acceptable arrangements of members can be used.

Arrangement #1
Space 8 x 8 crossbraces at six feet horizontally and five feet vertically.
Space 10 x 12 wales at five feet vertically.
Position 2 x 6 uprights as closely together as possible.
If water must be retained use special tongue and groove uprights to form tight sheeting.

Arrangement #2
Space 8 x 10 crossbraces at eight feet horizontally and five feet vertically.

Space 12 x 12 wales at five feet vertically.

Position 2 x 6 uprights in a close sheeting configuration unless water pressure must be resisted. Tight sheeting must be used where water must be retained.

(f)(4) Example 4

A trench dug in Type C soil is 20 feet deep and 11 feet wide. The size and spacing of members for the section of trench that is over 15 feet in depth is determined using Table C-1.3. Only one arrangement of members is provided.

Space 8 x 10 crossbraces at six feet horizontally and five feet vertically.

Space 12 x 12 wales at five feet vertically.

Use 3 x 6 tight sheeting.

Use of Tables C-2.1 through C-2.3 would follow the same procedures.

(g)

Notes for all Tables.

1. Member sizes at spacings other than indicated are to be determined as specified in 1926.652(c), "Design of Protective Systems."
2. When conditions are saturated or submerged use Tight Sheeting. Tight Sheeting refers to the use of specially-edged timber planks (e.g., tongue and groove) at least three inches thick, steel sheet piling, or similar construction that when driven or placed in position provide a tight wall to resist the lateral pressure of water and to prevent the loss of backfill material. Close Sheeting refers to the placement of planks side-by-side allowing as little space as possible between them.
3. All spacing indicated is measured center to center.
4. Wales to be installed with greater dimension horizontal.
5. f the vertical distance from the center of the lowest crossbrace to the bottom of the trench exceeds two and one-half feet, uprights shall be firmly embedded or a mudsill shall be used. Where uprights are embedded, the vertical distance from the center of the lowest crossbrace to the bottom of the trench shall not exceed 36 inches. When mudsills are used, the vertical distance shall not exceed 42 inches. Mudsills are wales that are installed at the tow of the trench side.
6. Trench jacks may be used in lieu of or in combination with timber crossbraces.
7. Placement of crossbraces. When the vertical spacing of crossbraces is four feet, place the top crossbrace no more than two feet below the top of the trench. When the vertical spacing of crossbraces is five feet, place the top crossbrace no more than 2.5 feet below the top of the trench.

TABLE C-1.1

TIMBER TRENCH SHORING -- MINIMUM TIMBER REQUIREMENTS *

SOIL TYPE A   P(a) = 25 X H + 72 psf (2 ft Surcharge)

DEPTH OF TRENCH (FEET)	HORIZ. SPACING (FEET)	SIZE (ACTUAL) AND SPACING OF MEMBERS **					VERT. SPACING (FEET)
		CROSS BRACES WIDTH OF TRENCH (FEET)					
		UP TO 4	UP TO 6	UP TO 9	UP TO 12	UP TO 15	
5 TO 10	UP TO 6	4X4	4X4	4X6	6X6	6X6	4
	UP TO 8	4X4	4X4	4X6	6X6	6X6	4
	UP TO 10	4X6	4X6	4X6	6X6	6X6	4
	UP TO 12	4X6	4X6	6X6	6X6	6X6	4
10 TO 15	UP TO 6	4X4	4X4	4X6	6X6	6X6	4
	UP TO 8	4X6	4X6	6X6	6X6	6X6	4
	UP TO 10	6X6	6X6	6X6	6X8	6X8	4
	UP TO 12	6X6	6X6	6X6	6X8	6X8	4
15 TO 20	UP TO 6	6X6	6X6	6X6	6X8	6X8	4
	UP TO 8	6X6	6X6	6X6	6X8	6X8	4
	UP TO 10	8X8	8X8	8X8	8X8	8X10	4
	UP TO 12	8X8	8X8	8X8	8X8	8X10	4
OVER 20	SEE NOTE 1							

TABLE C-1.1

TIMBER TRENCH SHORING -- MINIMUM TIMBER REQUIREMENTS *

SOIL TYPE A   P(a) = 25 X H + 72 psf (2 ft Surcharge)

DEPTH OF TRENCH (FEET)	WALES SIZE (IN)	WALES VERT. SPACING (FEET)	UPRIGHTS — MAXIMUM ALLOWABLE HORIZONTAL SPACING (FEET)				
			CLOSE	4	5	6	8
5 TO 10	Not Req'd	---				2X6	
	Not Req'd	---					2X8
	8X8	4			2X6		
	8X8	4				2X6	
10 TO 15	Not Req'd	---				3X8	
	8X8	4		2X6			
	8X10	4			2X6		
	10X10	4				3X8	
15 TO 20	6X8	4	3X6				
	8X8	4	3X6				
	8X10	4	3X6				
	10X10	4	3X6				
OVER 20	SEE NOTE 1						

* Mixed oak or equivalent with a bending strength not less than 850 psi.
** Manufactured members of equivalent strength may be substituted for wood.

TABLE C-1.2

TIMBER TRENCH SHORING -- MINIMUM TIMBER REQUIREMENTS *

SOIL TYPE B P(a) = 45 X H + 72 psf (2 ft Surcharge)

DEPTH OF TRENCH (FEET)	HORIZ. SPACING (FEET)	SIZE (ACTUAL) AND SPACING OF MEMBERS ** CROSS BRACES WIDTH OF TRENCH (FEET)					VERT. SPACING (FEET)
		UP TO 4	UP TO 6	UP TO 9	UP TO 12	UP TO 15	
5 TO 10	UP TO 6	4X6	4X6	6X6	6X6	6X6	5
	UP TO 8	6X6	6X6	6X6	6X8	6X8	5
	UP TO 10	6X6	6X6	6X6	6X8	6X8	5
	See Note 1						
10 TO 15	UP TO 6	6X6	6X6	6X6	6X8	6X8	5
	UP TO 8	6X8	6X8	6X8	8X8	8X8	5
	UP TO 10	8X8	8X8	8X8	8X8	8X10	5
	See Note 1						
15 TO 20	UP TO 6	6X8	6X8	6X8	8X8	8X8	5
	UP TO 8	8X8	8X8	8X8	8X8	8X10	5
	UP TO 10	8X10	8X10	8X10	8X10	10X10	5
	See Note 1						
OVER 20	SEE NOTE 1						

TABLE C-1.2

TIMBER TRENCH SHORING -- MINIMUM TIMBER REQUIREMENTS *

SOIL TYPE B $P(a) = 45 \times H + 72$ psf (2 ft Surcharge)

[Continued]

DEPTH OF TRENCH (FEET)	SIZE (ACTUAL) AND SPACING OF MEMBERS **					
	WALES		UPRIGHTS			
	SIZE (IN)	VERT. SPACING (FEET)	MAXIMUM ALLOWABLE HORIZONTAL SPACING (FEET)			
			CLOSE	2	3	
5 TO 10	6X8	5			2X6	
	8X10	5			2X6	
	10X10	5			2X6	
10 TO 15	8X8	5		2X6		
	10X10	5		2X6		
	10X12	5		2X6		
15 TO 20	8X10	5	3X6			
	10X12	5	3X6			
	12X12	5	3X6			
OVER 20	SEE NOTE 1					

* Mixed oak or equivalent with a bending strength not less than 850 psi.
** Manufactured members of equivalent strength may be substituted for wood.

TABLE C-1.3

TIMBER TRENCH SHORING -- MINIMUM TIMBER REQUIREMENTS *

SOIL TYPE C   P(a) = 80 X H + 72 psf (2 ft Surcharge)

DEPTH OF TRENCH (FEET)	HORIZ. SPACING (FEET)	SIZE (ACTUAL) AND SPACING OF MEMBERS ** CROSS BRACES WIDTH OF TRENCH (FEET)					VERT. SPACING (FEET)
		UP TO 4	UP TO 6	UP TO 9	UP TO 12	UP TO 15	
5 TO 10	UP TO 6	6X8	6X8	6X8	8X8	8X8	5
	UP TO 8	8X8	8X8	8X8	8X8	8X10	5
	UP TO 10	8X10	8X10	8X10	8X10	10X10	5
	See Note 1						
10 TO 15	UP TO 6	8X8	8X8	8X8	8X8	8X10	5
	UP TO 8	8X10	8X10	8X10	8X10	10X10	5
	See Note 1						
	See Note 1						
15 TO 20	UP TO 6	8X10	8X10	8X10	8X10	10X10	5
	See Note 1						
	See Note 1						
	See Note 1						
OVER 20	SEE NOTE 1						

TABLE C-1.3

TIMBER TRENCH SHORING -- MINIMUM TIMBER REQUIREMENTS *

SOIL TYPE C   P(a) = 80 X H + 72 psf (2 ft Surcharge)

[Continued]

DEPTH OF TRENCH (FEET)	SIZE (ACTUAL) AND SPACING OF MEMBERS **								
	WALES		UPRIGHTS						
	SIZE (IN)	VERT. SPACING (FEET)	MAXIMUM ALLOWABLE HORIZONTAL SPACING (FEET)					
			CLOSE						
5 TO 10	8X10	5	2X6						
	10X12	5	2X6						
	12X12	5	2X6						
10 TO 15	10X12	5	2X6						
	12X12	5	2X6						
15 TO 20	12X12	5	3X6						
OVER 20	SEE NOTE 1								

* Mixed oak or equivalent with a bending strength not less than 850 psi.
** Manufactured members of equivalent strength may be substituted for wood.

TABLE C-2.1

TIMBER TRENCH SHORING -- MINIMUM TIMBER REQUIREMENTS *

SOIL TYPE A $P(a) = 25 \times H + 72$ psf (2 ft Surcharge)

DEPTH OF TRENCH (FEET)	SIZE (S4S) AND SPACING OF MEMBERS **						
	CROSS BRACES						
	HORIZ. SPACING (FEET)	WIDTH OF TRENCH (FEET)					VERT. SPACING (FEET)
		UP TO 4	UP TO 6	UP TO 9	UP TO 12	UP TO 15	
5 TO 10	UP TO 6	4X4	4X4	4X4	4X4	4X6	4
	UP TO 8	4X4	4X4	4X4	4X6	4X6	4
	UP TO 10	4X6	4X6	4X6	6X6	6X6	4
	UP TO 12	4X6	4X6	4X6	6X6	6X6	4
10 TO 15	UP TO 6	4X4	4X4	4X4	6X6	6X6	4
	UP TO 8	4X6	4X6	4X6	6X6	6X6	4
	UP TO 10	6X6	6X6	6X6	6X6	6X6	4
	UP TO 12	6X6	6X6	6X6	6X6	6X6	4
15 TO 20	UP TO 6	6X6	6X6	6X6	6X6	6X6	4
	UP TO 8	6X6	6X6	6X6	6X6	6X6	4
	UP TO 10	6X6	6X6	6X6	6X6	6X8	4
	UP TO 12	6X6	6X6	6X6	6X8	6X8	4
OVER 20	SEE NOTE 1						

TABLE C-2.1

TIMBER TRENCH SHORING -- MINIMUM TIMBER REQUIREMENTS *

SOIL TYPE A   P(a) = 25 X H + 72 psf (2 ft Surcharge)

[Continued]

DEPTH OF TRENCH (FEET)	SIZE (S4S) AND SPACING OF MEMBERS **						
	WALES		UPRIGHTS				
	SIZE (IN)	VERT. SPACING (FEET)	MAXIMUM ALLOWABLE HORIZONTAL SPACING (FEET)				
			CLOSE	4	5	6	8
5 TO 10	Not Req'd	Not Req'd				4X6	
	Not Req'd	Not Req'd					4X8
	8X8	4			4X6		
	8X8	4				4X6	
10 TO 15	Not Req'd	Not Req'd				4X10	
	6X8	4		4X6			
	8X8	4			4X8		
	8X10	4		4X6		4X10	
15 TO 20	6X8	4	3X6				
	8X8	4	3X6	4x12			
	8X10	4	3X6				
	8X12	4	3X6	4x12			
OVER 20	SEE NOTE 1						

* Douglas fir or equivalent with a bending strength not less than 1500 psi.
** Manufactured members of equivalent strength may be substituted for wood.

TABLE C-2.2

TIMBER TRENCH SHORING -- MINIMUM TIMBER REQUIREMENTS *

SOIL TYPE B P(a) = 45 X H + 72 psf (2 ft Surcharge)

DEPTH OF TRENCH (FEET)	HORIZ. SPACING (FEET)	\multicolumn{5}{c}{WIDTH OF TRENCH (FEET)}	VERT. SPACING (FEET)
		UP TO 4	UP TO 6	UP TO 9	UP TO 12	UP TO 15	
5 TO 10	UP TO 6	4X6	4X6	4X6	6X6	6X6	5
	UP TO 8	4X6	4X6	6X6	6X6	6X6	5
	UP TO 10	4X6	4X6	6X6	6X6	6X8	5
	See Note 1						
10 TO 15	UP TO 6	6X6	6X6	6X6	6X8	6X8	5
	UP TO 8	6X8	6X8	6X8	8X8	8X8	5
	UP TO 10	6X8	6X8	8X8	8X8	8X8	5
	See Note 1						
15 TO 20	UP TO 6	6X8	6X8	6X8	6X8	8X8	5
	UP TO 8	6X8	6X8	6X8	8X8	8X8	5
	UP TO 10	8X8	8X8	8X8	8X8	8X8	5
	See Note 1						
OVER 20	SEE NOTE 1						

SIZE (S4S) AND SPACING OF MEMBERS ** — CROSS BRACES

TABLE C-2.2

TIMBER TRENCH SHORING -- MINIMUM TIMBER REQUIREMENTS *

SOIL TYPE B P(a) = 45 X H + 72 psf (2 ft Surcharge)

[Continued]

DEPTH OF TRENCH (FEET)	SIZE (S4S) AND SPACING OF MEMBERS **						
	WALES		UPRIGHTS				
	SIZE (IN)	VERT. SPACING (FEET)	MAXIMUM ALLOWABLE HORIZONTAL SPACING (FEET)				
			CLOSE	2	3		
5 TO 10	6X8	5			3X12 4X8		4X12
	8X8	5		3X8		4X8	
	8X10	5			4X8		
10 TO 15	8X8	5	3X6	4X10			
	10X10	5	3X6	4X10			
	10X12	5	3X6	4X10			
15 TO 20	8X10	5	4X6				
	10X12	5	4X6				
	12X12	5	4X6				
OVER 20	SEE NOTE 1						

* Douglas fir or equivalent with a bending strength not less than 1500 psi.
** Manufactured members of equivalent strength may be substituted for wood.

TABLE C-2.3

TIMBER TRENCH SHORING -- MINIMUM TIMBER REQUIREMENTS *

SOIL TYPE C   P(a) = 80 X H + 72 psf (2 ft Surcharge)

DEPTH OF TRENCH (FEET)	HORIZ. SPACING (FEET)	\multicolumn{5}{c}{CROSS BRACES — WIDTH OF TRENCH (FEET)}	VERT. SPACING (FEET)
		UP TO 4	UP TO 6	UP TO 9	UP TO 12	UP TO 15	
5 TO 10	UP TO 6	6X6	6X6	6X6	6X6	8X8	5
	UP TO 8	6X6	6X6	6X6	8X8	8X8	5
	UP TO 10	6X6	6X6	8X8	8X8	8X8	5
	See Note 1						
10 TO 15	UP TO 6	6X8	6X8	6X8	8X8	8X8	5
	UP TO 8	8X8	8X8	8X8	8X8	8X8	5
	See Note 1						
	See Note 1						
15 TO 20	UP TO 6	8X8	8X8	8X8	8X10	8X10	5
	See Note 1						
	See Note 1						
	See Note 1						
OVER 20	SEE NOTE 1						

TABLE C-2.3

TIMBER TRENCH SHORING -- MINIMUM TIMBER REQUIREMENTS *

SOIL TYPE C   P(a) = 80 X H + 72 psf (2 ft Surcharge)

[Continued]

DEPTH OF TRENCH (FEET)	SIZE (S4S) AND SPACING OF MEMBERS **						
	WALES		UPRIGHTS				
	SIZE (IN)	VERT. SPACING (FEET)	MAXIMUM ALLOWABLE HORIZONTAL SPACING (FEET)				
			CLOSE				
5 TO 10	8X8	5	3X6				
	10X10	5	3X6				
10	10X12	5	3X6				
10 TO 15	10X10	5	4X6				
	12X12	5	4X6				
15 TO 20	10X12	5	4X6				
OVER 20	SEE NOTE 1						

* Douglas fir or equivalent with a bending strength not less than 1500 psi.
** Manufactured members of equivalent strength may be substituted for wood.

ALUMINUM HYDRAULIC SHORING FOR TRENCHES—1926 SUBPART P APP D

(a)

Scope. This appendix contains information that can be used when aluminum hydraulic shoring is provided as a method of protection against cave-ins in trenches that do not exceed 20 feet (6.1m) in depth. This appendix must be used when design of the aluminum hydraulic protective system cannot be performed in accordance with 1926.652(c)(2).

(b)

Soil Classification. In order to use data presented in this appendix, the soil type or types in which the excavation is made must first be determined using the soil classification method set forth in appendix A of subpart P of part 1926.

(c)

Presentation of Information. Information is presented in several forms as follows:

(c)(1)

Information is presented in tabular form in Tables D-1.1, D-1.2, D-1.3 and D-1.4. Each table presents the maximum vertical and horizontal spacings that may be used with various aluminum member sizes and various hydraulic cylinder sizes. Each table contains data only for the particular soil type in which the excavation or portion of the excavation is made. Tables D-1.1 and D-1.2 are for vertical shores in Types A and B soil. Tables D-1.3 and D-1.4 are for horizontal waler systems in Types B and C soil.

(c)(2)

Information concerning the basis of the tabular data and the limitations of the data is presented in paragraph (d) of this appendix.

(c)(3)

Information explaining the use of the tabular data is presented in paragraph (e) of this appendix.

(c)(4)

Information illustrating the use of the tabular data is presented in paragraph (f) of this appendix.

(c)(5)

Miscellaneous notations (Footnotes) regarding Table D-1.1 through D-1.4 are presented in paragraph (g) of this appendix.

(c)(6)

Figures, illustrating typical installations of hydraulic shoring, are included just prior to the Tables. The illustrations page is entitled "Aluminum Hydraulic Shoring: Typical Installations."

(d)

Basis and limitations of the data.

(d)(1)

Vertical shore rails and horizontal wales are those that meet the Section Modulus requirements in the D-1 Tables. Aluminum material is 6061-T6 or material of equivalent strength and properties.

(d)(2)

Hydraulic cylinders specifications.

(d)(2)(i)

Two-inch cylinders shall be a minimum 2-inch inside diameter with a minimum safe working capacity of no less than 18,000 pounds axial compressive load at maximum extension. Maximum extension is to include full range of cylinder extensions as recommended by product manufacturer.

(d)(2)(ii)

3-inch cylinders shall be a minimum 3-inch inside diameter with a safe working capacity of not less than 30,000 pounds axial compressive load at extensions as recommended by product manufacturer.

(d)(3)

Limitation of application.

(d)(3)(i)

It is not intended that the aluminum hydraulic specification apply to every situation that may be experienced in the field. These data were developed to apply to the situations that are most commonly experienced in current trenching practice. Shoring systems for use in situations that are not covered by the data in this appendix must be otherwise designed as specified in 1926.652(c).

(d)(3)(ii)

When any of the following conditions are present, the members specified in the Tables are not considered adequate. In this case, an alternative aluminum hydraulic shoring system or other type of protective system must be designed in accordance with 1926.652.

(d)(3)(ii)(A)

When vertical loads imposed on cross braces exceed a 100 Pound gravity load distributed on a one foot section of the center of the hydraulic cylinder.

(d)(3)(ii)(B)

When surcharge loads are present from equipment weighing in excess of 20,000 pounds.

(d)(3)(ii)(C)

When only the lower portion of a trench is shored and the remaining portion of the trench is sloped or benched unless: The sloped portion is sloped at an angle less steep than three horizontal to one vertical; or the members are selected from the tables for use at a depth which is determined from the top of the overall trench, and not from the toe of the sloped portion.

(e)

Use of Tables D-1.1, D-1.2, D-1.3 and D-1.4. The members of the shoring system that are to be selected using this information are the hydraulic cylinders, and either the vertical shores or the horizontal wales. When a waler system is used the vertical timber sheeting to be used is also selected from these tables. The Tables

D-1.1 and D-1.2 for vertical shores are used in Type A and B soils that do not require sheeting. Type B soils that may require sheeting, and Type C soils that always require sheeting, are found in the horizontal wale Tables D-1.3 and D-1.4. The soil type must first be determined in accordance with the soil classification system described in appendix A to subpart P of part 1926. Using the appropriate table, the selection of the size and spacing of the members is made. The selection is based on the depth and width of the trench where the members are to be installed. In these tables the vertical spacing is held constant at four feet on center. The tables show the maximum horizontal spacing of cylinders allowed for each size of wale in the waler system tables, and in the vertical shore tables, the hydraulic cylinder horizontal spacing is the same as the vertical shore spacing.

(f)

Example to illustrate the use of the tables:

(f)(1) Example 1

A trench dug in Type A soil is 6 feet deep and 3 feet wide. From Table D-1.1: Find vertical shores and 2 inch diameter cylinders spaced 8 feet on center (o.c.) horizontally and 4 feet on center (o.c.) vertically. (See Figures 1 and 3 for typical installations.)

(f)(2) Example 2

A trench is dug in Type B soil that does not require sheeting, 13 feet deep and 5 feet wide. From Table D-1.2: Find vertical shores and 2 inch diameter cylinders spaced 6.5 feet o.c. horizontally and 4 feet o.c. vertically. (See Figures 1 and 3 for typical installations.)

(f)(3) Example 3

A trench is dug in Type B soil that does not require sheeting, but does experience some minor raveling of the trench face. the trench is 16 feet deep and 9 feet wide. From Table D-1.2: Find vertical shores and 2 inch diameter cylinder (with special oversleeves as designated by Footnote #2) spaced 5.5 feet o.c. horizontally and 4 feet o.c. vertically. Plywood (per Footnote (g)(7) to the D-1 Table) should be used behind the shores. (See Figures 2 and 3 for typical installations.)

(f)(4) Example 4

A trench is dug in previously disturbed Type B soil, with characteristics of a Type C soil, and will require sheeting. The trench is 18 feet deep, and 12 feet wide 8 foot horizontal spacing between cylinders is desired for working space. From Table D-1.3: Find horizontal wale with a section modulus of 14.0 spaced at 4 feet o.c. vertically and 3 inch diameter cylinder spaced at 9 feet maximum o.c. horizontally, 3 x 12 timber sheeting is required at close spacing vertically. (See Figure 4 for typical installation.)

(f)(5) Example 5

A trench is dug in Type C soil, 9 feet deep and 4 feet wide. Horizontal cylinder spacing in excess of 6 feet is desired for working space. From Table D-1.4: Find horizontal wale with a section modulus of 7.0 and 2 inch diameter cylinders spaced at 6.5 feet o.c. horizontally. Or, find horizontal wale with a 14.0 section modulus and 3 inch diameter cylinder spaced at 10 feet o.c. horizontally. Both wales are spaced 4 feet o.c. vertically, 3 x 12 timber sheeting is required at close spacing vertically. (See Figure 4 for typical installation.)

(g)

Footnotes, and general notes, for Tables D-1.1, D-1.2, D-1.3, and D-1.4.

(g)(1)

For applications other than those listed in the tables, refer to 1926.652(c)(2) for use of manufacturer's tabulated data. For trench depths in excess of 20 feet, refer to 1926.652(c)(2) and 1926.652(c)(3).

(g)(2)

Two-inch diameter cylinders, at this width, shall have structural steel tube (3.5 x 3.5 x 0.1875) oversleeves, or structural oversleeves of manufacturer's specification, extending the full, collapsed length.

(g)(3)

Hydraulic cylinders capacities.

(g)(3)(i)

Two-inch cylinders shall be a minimum 2-inch inside diameter with a safe working capacity of not less than 18,000 pounds axial compressive load at maximum extension. Maximum extension is to include full range of cylinder extensions as recommended by product manufacturer.

(g)(3)(ii)

Three-inch cylinders shall be a minimum 3-inch inside diameter with a safe work capacity of not less than 30,000 pounds axial compressive load at maximum extension. Maximum extension is to include full range of cylinder extensions as recommended by product manufacturer.

(g)(4)

All spacing indicated is measured center to center.

(g)(5)

Vertical shoring rails shall have a minimum section modulus of 0.40 inch.

(g)(6)

When vertical shores are used, there must be a minimum of three shores spaced equally, horizontally, in a group.

(g)(7)

Plywood shall be 1.125 inch thick softwood or 0.75 inch thick, 14 ply, arctic white birch (Finland form). Please note that plywood is not intended as a structural member, but only for prevention of local raveling (sloughing of the trench face) between shores.

(g)(8)

See appendix C for timber specifications.

(g)(9)

Wales are calculated for simple span conditions.

(g)(10)

See appendix D, item (d), for basis and limitations of the data.

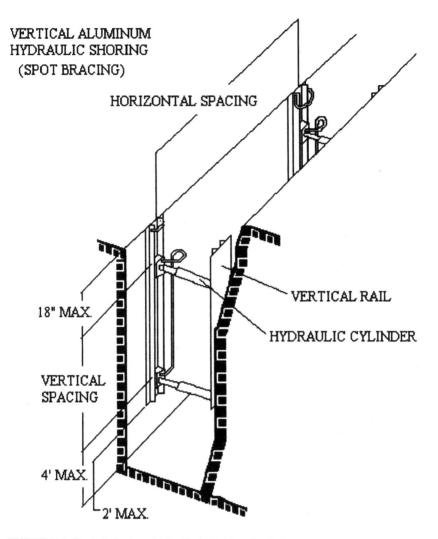

FIGURE D.1 Vertical aluminum hydraulic shoring (spot bracing).

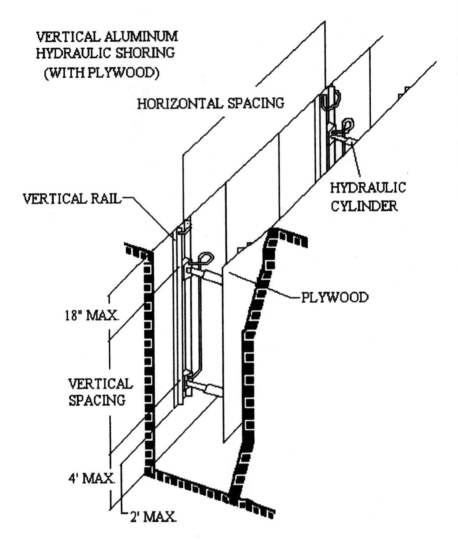

FIGURE D.2 Vertical aluminum hydraulic shoring (with plywood).

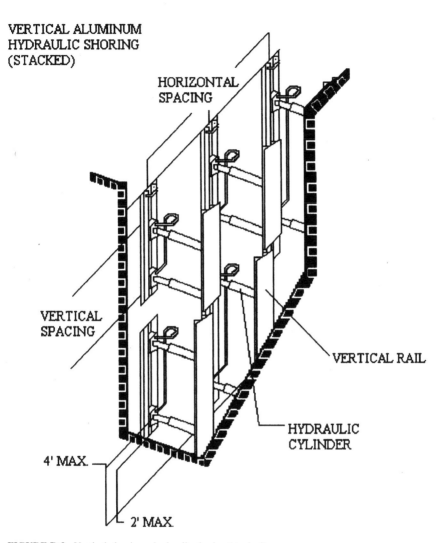

FIGURE D.3 Vertical aluminum hydraulic shoring (stacked).

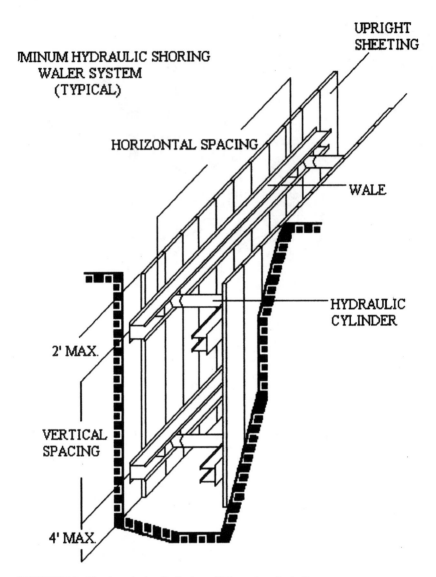

FIGURE D.4 Aluminum hydraulic shoring—Waler system (typical).

TABLE D - 1.1
ALUMINUM HYDRAULIC SHORING
VERTICAL SHORES
FOR SOIL TYPE A

DEPTH OF TRENCH (FEET)	HYDRAULIC CYLINDERS				
	MAXIMUM HORIZONTAL SPACING (FEET)	MAXIMUM VERTICAL SPACING (FEET)	WIDTH OF TRENCH (FEET)		
			UP TO 8	OVER 8 UP TO 12	OVER 12 UP TO 15
OVER 5 UP TO 10	8				
OVER 10 UP TO 15	8	4	2 INCH DIAMETER	2 INCH DIAMETER NOTE(2)	3 INCH DIAMETER
OVER 15 UP TO 20	7				
OVER 20	NOTE(1)				

Footnotes to tables, and general notes on hydraulic shoring, are found in Appendix D, Item (g)
Note(1): See Appendix D, Item (g)(1)
Note(2): See Appendix D, Item (g)(2)

FIGURE D.1.1 Aluminum hydraulic shoring vertical shores for soil type A.

TABLE D - 1.2
ALUMINUM HYDRAULIC SHORING
VERTICAL SHORES
FOR SOIL TYPE B

DEPTH OF TRENCH (FEET)	MAXIMUM HORIZONTAL SPACING (FEET)	HYDRAULIC CYLINDERS			
		MAXIMUM VERTICAL SPACING (FEET)	WIDTH OF TRENCH (FEET)		
			UP TO 8	OVER 8 UP TO 12	OVER 12 UP TO 15
OVER 5 UP TO 10	8				
OVER 10 UP TO 15	6.5	4	2 INCH DIAMETER	2 INCH DIAMETER NOTE(2)	3 INCH DIAMETER
OVER 15 UP TO 20	5.5				
OVER 20	NOTE(1)				

Footnotes to tables, and general notes on hydraulic shoring, are found in Appendix D, Item (g)
Note(1): See Appendix D, Item (g)(1)
Note(2): See Appendix D, Item (g)(2)

FIGURE D.1.2 Aluminum hydraulic shoring vertical shores for soil type B.

ALUMINUM HYDRAULIC SHORING
WALER SYSTEMS
FOR SOIL TYPE B

DEPTH OF TRENCH (FEET)	WALES VERTICAL SPACING (FEET)	* SECTION MODULUS (IN(3))	HYDRAULIC CYLINDERS			
			WIDTH OF TRENCH (FEET)			
			UP TO 8		OVER 8 UP TO 12	
			HORIZ SPACING	CYLINDER DIAMETER	HORIZ SPACING	CYLINDER DIAMETER
OVER 5 UP TO 10	4	3.5	8.0	2 IN	8.0	2 IN NOTE (2)
		7.0	9.0	2 IN	9.0	2 IN NOTE (2)
		14.0	12.0	3 IN	12.0	3 IN
OVER 10 UP TO 15	4	3.5	6.0	2 IN	6.0	2 IN NOTE (2)
		7.0	8.0	3 IN	8.0	3 IN
		14.0	10.0	3 IN	10.0	3 IN
OVER 15 UP TO 20	4	3.5	5.5	2 IN	5.5	2 IN NOTE (2)
		7.0	6.0	3 IN	6.0	3 IN
		14.0	9.0	3 IN	9.0	3 IN
OVER 20	NOTE (1)					

FIGURE D.1.3 Aluminum hydraulic shoring waler systems for soil type B.

TABLE D - 1.3
ALUMINUM HYDRAULIC SHORING
WALER SYSTEMS
FOR SOIL TYPE B

[Continued]

DEPTH OF TRENCH (FEET)	WALES		HYDRAULIC CYLINDERS		TIMBER UPRIGHTS		
	VERTICAL SPACING (FEET)	* SECTION MODULUS (IN(3))	WIDTH OF TRENCH (FEET) OVER 12 UP TO 15		MAX. HORIZ SPACING (ON CENTER)		
			HORIZ SPACING	CYLINDER DIAMETER	SOLID SHEET	2 FT	3 FT
OVER 5 UP TO 10	4	3.5	8.0	3 IN			
		7.0	9.0	3 IN	---	---	3x12
		14.0	12.0	3 IN			
OVER 10 UP TO 15	4	3.5	6.0	3 IN			
		7.0	8.0	3 IN	---	3x12	---
		14.0	10.0	3 IN			
OVER 15 UP TO 20	4	3.5	5.5	3 IN			
		7.0	6.0	3 IN	3x12	---	---
		14.0	9.0	3 IN			
OVER 20		NOTE (1)					

Footnotes to tables, and general notes on hydraulic shoring, are found in Appendix D, Item (g)
Note(1): See Appendix D, Item (g)(1)
Note(2): See Appendix D, Item (g)(2)
* Consult product manufacturer and/or qualified engineer for Section Modulus of available wales.

FIGURE D.1.3 *(continued)* Aluminum hydraulic shoring waler systems shores for soil type B.

TABLE D - 1.4
ALUMINUM HYDRAULIC SHORING
WALER SYSTEMS
FOR SOIL TYPE C

DEPTH OF TRENCH (FEET)	WALES		HYDRAULIC CYLINDERS			
	VERTICAL SPACING (FEET)	* SECTION MODULUS (IN(3))	WIDTH OF TRENCH (FEET)			
			UP TO 8		OVER 8 UP TO 12	
			HORIZ SPACING	CYLINDER DIAMETER	HORIZ SPACING	CYLINDER DIAMETER
OVER 5 UP TO 10	4	3.5	6.0	2 IN	6.0	2 IN NOTE (2)
		7.0	6.5	2 IN	6.5	2 IN NOTE (2)
		14.0	10.0	3 IN	10.0	3 IN
OVER 10 UP TO 15	4	3.5	4.0	2 IN	4.0	2 IN NOTE (2)
		7.0	5.5	3 IN	5.5	3 IN
		14.0	8.0	3 IN	8.0	3 IN
OVER 15 UP TO 20	4	3.5	3.5	2 IN	3.5	2 IN NOTE (2)
		7.0	5.0	3 IN	5.0	3 IN
		14.0	6.0	3 IN	6.0	3 IN
OVER 20		NOTE (1)				

FIGURE D.1.4 Aluminum hydraulic shoring waler systems for soil type C.

TABLE D - 1.4
ALUMINUM HYDRAULIC SHORING
WALER SYSTEMS
FOR SOIL TYPE C

[Continued]

DEPTH OF TRENCH (FEET)	WALES		HYDRAULIC CYLINDERS		TIMBER UPRIGHTS		
	VERTICAL SPACING (FEET)	* SECTION MODULUS (IN(3))	WIDTH OF TRENCH (FEET) OVER 12 UP TO 15		MAX. HORIZ SPACING (ON CENTER)		
			HORIZ SPACING	CYLINDER DIAMETER	SOLID SHEET	2 FT	3 FT
OVER 5 UP TO 10	4	3.5	6.0	3 IN			
		7.0	6.5	3 IN	3x12	---	---
		14.0	10.0	3 IN			
OVER 10 UP TO 15	4	3.5	4.0	3 IN			
		7.0	5.5	3 IN	3x12	---	---
		14.0	8.0	3 IN			
OVER 15 UP TO 20	4	3.5	3.5	3 IN			
		7.0	5.0	3 IN	3x12	---	---
		14.0	6.0	3 IN			
OVER 20		NOTE (1)					

Footnotes to tables, and general notes on hydraulic shoring, are found in Appendix D, Item (g)
Note(1): See Appendix D, Item (g)(1)
Note(2): See Appendix D, Item (g)(2)
* Consult product manufacturer and/or qualified engineer for Section Modulus of available wales.

FIGURE D.1.4 *(continued)* Aluminum hydraulic waler systems for soil type C.

ALTERNATIVES FOR TIMBER SHORING—
1926 SUBPART P APPE

Standard Number: 1926SubpartPAppE
Standard Title: Alternatives to Timber Shoring
SubPart Number: P
SubPart Title: Excavations

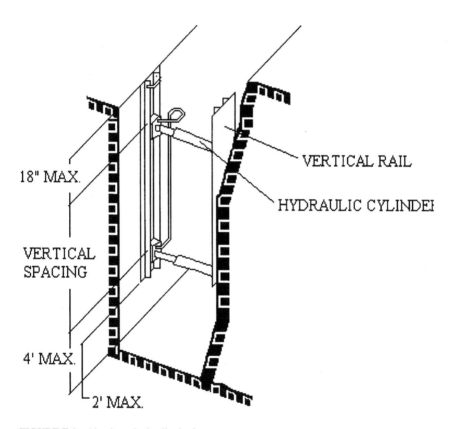

FIGURE E.1 Aluminum hydraulic shoring.

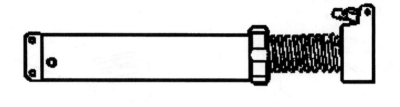

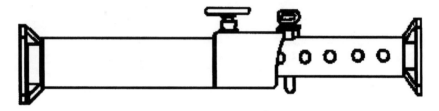

FIGURE E.2 Pneumatic/hydraulic shoring.

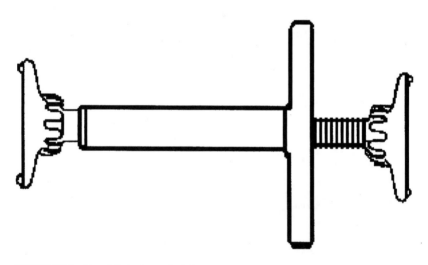

FIGURE E.3 Trench jacks (screw jacks).

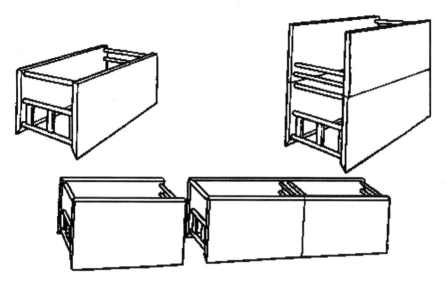

FIGURE E.4 Trench shields.

SELECTION OF PROTECTIVE SYSTEMS—1926 SUBPART P APP F

The following figures are a graphic summary of the requirements contained in subpart P for excavations 20 feet or less in depth. Protective systems for use in excavations more than 20 feet in depth must be designed by a registered professional engineer in accordance with 1926.652(b) and (c).

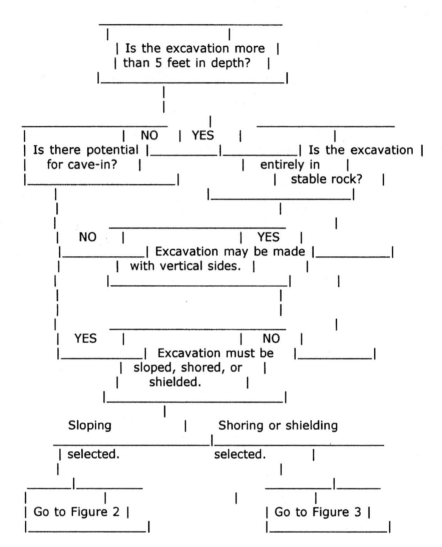

FIGURE F.1 Preliminary decisions.

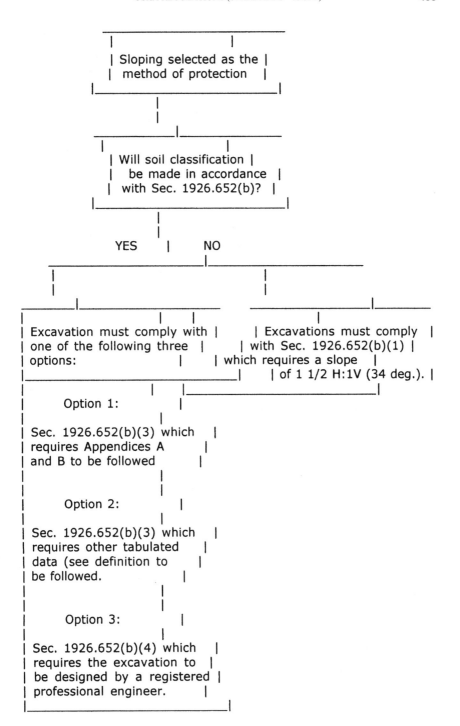

FIGURE F.2 Sloping options.

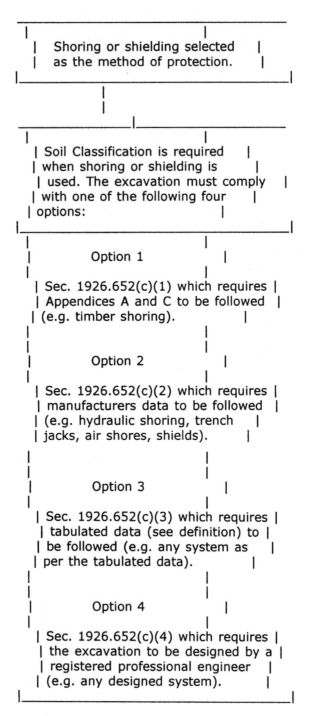

FIGURE F.3 Shoring and shielding options.

CHAPTER 17
COMPUTER DESIGN

This chapter discusses two sample computer programs which are used to design shoring systems. The first program, called ct-Shoring, is used to design soldier pile and lagging, sheet piling, and secant and cylinder pile walls, and utilizes analysis methods from earth pressure theory and apparent earth pressure. The input data required is discussed in the preceding chapters of this book.

The second program, called GoldNail, is used for soil nail designs and utilizes limit equilibrium analysis. The input data required is also discussed in the previous chapters of this book.

The user of these programs is cautioned that the use of computer solutions without experience with these types of calculations should not be undertaken unless the user is prepared to do manual check calculations to ensure that the answers are rational. Methods of performing check calculations are briefly discussed in Chapter 11 of this book and are available in much more detail from listings in the Bibliography.

INTRODUCTION TO CIVILTECH'S SHORING SOFTWARE

ct-SHORING Suite Plus is a design and analysis software for excavation support systems, including braced cuts, raker support, trench box, cantilever walls, bulkhead walls, sheet pile walls, soldier piles and lagging systems, tangent pile walls, slurry walls, and any flexible walls. The program is flexible enough to handle any complex ground conditions, sloped surfaces, surcharge loads, and water conditions. Users have two choices to input data:

1. Input soil parameters such as friction angle, unit weight, and water condition, or
2. Input pressures such as active, passive, and hydraulic pressures.

The program calculates the embedment and maximum moment of the pile and automatically selects the sheet pile and soldier pile from a database. It then gives properties and top deflection of selected piles. The program supports multi-tieback systems and calculates the free length, bond length, and no-load zone of tieback anchors.

The program presents diagrams of pressures, shear, moment, and deflection. The calculations are based on Federal Highway Administration (FHWA) methods, US Navy DM-7 (NAVFAC) manual, Caltrans Trenching & Shoring Manual, and the Steel Sheet Design (USS) manual. ct-SHORING Suite Plus includes four programs, which are linked together:

Epres determines the pressure diagram based on the soil and ground conditions.

Lpres determines the lateral pressures due to surcharge such as point, line, strip, and area loads.

Heave checks the overall stability of the shoring system in soft ground.

Shoring performs the shoring wall calculation and analysis based on the input pressures.

FEATURES

- Windows based, user-friendly interface
- Easy installation, easy to learn and easy to use
- Unlimited layered soils below and above dredge line
- Step-by-step manual with 25 examples
- Up to 20 levels of tieback or braces
- Diagrams of pressures, shear, moment, and deflection
- Selection of piles, determination of tieback length
- Deflection for the selected piles
- Optimized pile size selection from a database
- Input different spacing for each pressure
- Sloped ground surface
- Up to 20 types of surcharge
- Accommodates different water tables and seepage or non-seepage conditions at pile tip
- High quality graphical reports that can be exported to other software

REQUIREMENTS

- IBM PC-compatible 486 or better
- Minimum system memory of 640K
- Windows NT, 95, 98, ME, 2000, or XP

GOLDNAIL

GoldNail version 3.11 is a 16-bit Windows-based soil nail design and analysis program. Golder Associates initially developed GoldNail to meet internal demand for a versatile, user-friendly design tool. At the encouragement of FHWA, Golder Associates made GoldNail available for sale to the general public.

THEORY

GoldNail is a slip surface limiting equilibrium model based on satisfying overall limiting equilibrium (translational and rotational) of individual free bodies defined by circular slip surfaces. For each nail intersecting the slip surface, the support provided by that nail is defined by the nail tension distribution that is characterized by the factored soil-nail adhesion, the factored nail strength, and the factored nail head strength. Unknowns are:

- Magnitude of normal force N
- Distribution of normal stress (or force) along the slip surface $N = N(1)$
- Factor of safety relating the shear strength to the normal stress $S = f(N) / FS$

The solution proceeds iteratively:

- For an assumed normal stress distribution $N = N(1)$, solve for the magnitude of the normal force and the factor of safety on soil strength, using the two equations of translational equilibrium.
- Check for moment equilibrium. If resisting moments equal disturbing moments, solution is obtained.
- If moments do not balance, modify the normal stress distribution to increase or decrease the resisting movement, as required, and repeat the process until moment balance is achieved.

The process can also be reversed to solve for a nail pattern (lengths and capacities) if a soil strength factor of safety is specified i.e., total required nail reinforcing force replaces the soil factor of safety as one of the unknowns. A pattern of nails (lengths and capacities) is then developed to provide the required nail reinforcing force for each slip surface considered.

ANALYSIS WITH NAILS/WITHOUT NAILS

You can choose to analyze a soil-nailed wall (analysis with nails) or perform a slope-stability analysis (analysis without nails) with the option to consider facing pressure on the slope. For information on GoldNail, or to purchase a copy, please contact:

Golder Associates Inc.
18300 NE Union Hill Road
Redmond, WA 98052
Tel: (425) 883-0777
Fax: (425) 882-5498 Attn: Joe Hachey
Email: jhachey@golder.com

CHAPTER 18
TABLES

This chapter consists of tables which are useful in the design and construction of soldier pile and lagging walls, tiedback walls, or soil nailed walls. Applications to secants walls, slurry walls, and cylinder walls are also appropriate.

Figure 18.1 is a table which indicates the dimensions, weights, and engineering properties of HP sections. HP sections are most frequently used in driven soldier piles or single beam/tieback through waler applications. See references in Chapters 3.5.1.1, 4.4.2.5, and 11.1-11.3.

Figure 18.2 is a table which indicates the dimensions, weights, and engineering properties of wide flange sections most often used in shoring applications. Wide flange sections are most frequently used in drilled and placed soldier piles, double beam/tieback through waler applications, rakers, and conventional walers. See references in Chapters 3.5.1.2, 3.5.1.4, 4.4.2.5, and 11.1-11.3.

Figure 18.3 is a table which indicates the dimensions, weights and engineering properties of channel sections often used in shoring applications. Channels are used in double channel soldier piles as well as double channel/tieback or deadman through waler applications. See references in Chapters 3.5.1.2, 4.4.2.5, and 11.1-11.4.

Figure 18.4 is a table which indicates the potential unit pullout capacity of various soils. In the absence of specific site information, these tables are useful to design tieback or soil nail lengths prior to site confirmation of the values used. It must be emphasized that these values are only first approximations of capacity and should only be used for purposes of obtaining initial estimates of the tieback or soil nail lengths. See references in Chapters 3.6, 4.4, 4.7, 11.5, and 11.9.

Figure 18.5 is a table which can be used to calculate the neat volume of grout required for tieback, soil nail and micropile installation. See References in Chapters 3.6, 3.10, 4.4, 4.7, 11.5, and 11.9

Figure 18.6 is a table which can be used to calculate the neat volume of concrete required for drilled pile installation. See References in Chapters 3.5, 3.7, 3.8, 3.11, 11.1- 11.3, and 11.6.

Figure 18.7 consists of sample soil nail recording forms which can be used for monitoring drilling and grouting of soil nails or tiebacks. See References in Chapter 3.6, 4.4, 4.7, 11.5, and 11.9.

Figure 18.8 involves tables of allowable timber stresses for design of lagging. See references in Chapter 3.5, 5.1, and 11.8.

Figure 18.9 is a set of mix designs for tieback grout, structural concrete for soldier pile toes, structural concrete for shotcrete, and Controlled Density Fill (CDF) lean mix for soldier piles.

Steel H-PILES
Imperial and *Metric*

SKYLINE STEEL CORPORATION

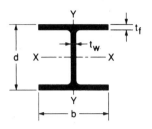

Dimensions and Properties

SECTION	WEIGHT lbs/ft kg/m	AREA in^2 cm^2	DEPTH d in mm	FLANGE Width b in mm	FLANGE Thick. t_f in mm	WEB THICK. t_w in mm	COATING AREA ft^2/ft m^2/m	ELASTIC PROPERTIES Axis X-X I in^4 cm^4	S in^3 cm^3	r in cm	Axis Y-Y I in^4 cm^4	S in^3 cm^3	r in cm
HP 14 / HP 360	117 / 174	34.6 / 222	14.21 / 361	14.885 / 378	0.805 / 20.4	0.805 / 20.4	7.20 / 2.19	1230 / 51100	173 / 2830	5.96 / 15.2	443 / 18300	59.5 / 968	3.58 / 9.08
	102 / 152	30.2 / 194	14.01 / 356	14.785 / 376	0.705 / 17.9	0.705 / 17.9	7.15 / 2.18	1060 / 44200	151 / 2480	5.92 / 15.1	380 / 15800	51.4 / 840	3.55 / 9.00
	89 / 132	26.3 / 169	13.83 / 351	14.695 / 373	0.615 / 15.6	0.615 / 15.6	7.10 / 2.16	913 / 37800	132 / 2150	5.89 / 15.0	326 / 13500	44.3 / 724	3.52 / 8.94
	73 / 108	21.6 / 138	13.61 / 346	14.585 / 370	0.505 / 12.8	0.505 / 12.8	7.05 / 2.15	738 / 30600	108 / 1770	5.84 / 14.8	261 / 10800	35.8 / 584	3.48 / 8.81
HP 12 / HP 310	84 / 125	24.6 / 158	12.28 / 312	12.295 / 312	0.685 / 17.4	0.685 / 17.4	6.03 / 1.84	650 / 27000	106 / 1730	5.14 / 13.1	213 / 8820	34.5 / 565	2.94 / 7.47
	74 / 110	21.8 / 140	12.13 / 308	12.215 / 310	0.610 / 15.5	0.605 / 15.4	5.99 / 1.83	569 / 23600	93.8 / 1530	5.11 / 13.0	186 / 7700	30.3 / 497	2.92 / 7.42
	63 / 93	18.4 / 118	11.94 / 303	12.125 / 308	0.515 / 13.1	0.515 / 13.1	5.95 / 1.81	472 / 19600	79.1 / 1290	5.07 / 12.9	153 / 6380	25.2 / 414	2.89 / 7.35
	53 / 79	15.5 / 99.7	11.78 / 299	12.045 / 306	0.435 / 11.0	0.435 / 11.0	5.91 / 1.80	393 / 16200	66.7 / 1080	5.04 / 12.7	127 / 5250	21.0 / 343	2.86 / 7.26
HP 10 / HP 250	57 / 85	16.7 / 108	9.99 / 254	10.225 / 260	0.565 / 14.4	0.565 / 14.4	4.98 / 1.52	294 / 12300	58.7 / 969	4.19 / 10.7	101 / 4220	19.7 / 325	2.46 / 6.25
	42 / 62	12.3 / 79.8	9.70 / 246	10.075 / 256	0.420 / 10.7	0.415 / 10.5	4.91 / 1.50	210 / 8750	43.3 / 711	4.14 / 10.5	71.7 / 2990	14.2 / 234	2.41 / 6.12
HP 8 / HP 200	36 / 53	10.5 / 68.1	8.02 / 204	8.155 / 207	0.445 / 11.3	0.445 / 11.3	3.98 / 1.21	119 / 4970	29.7 / 487	3.37 / 8.54	40.3 / 1670	9.88 / 161	1.96 / 4.95

FIGURE 18.1 H Piles. *(Courtesy of Skyline Steel, Inc. Gig Harbor, WA)*

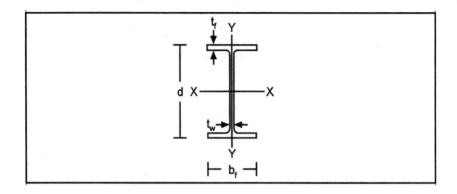

WIDE FLANGED BEAM TABLES

Designation	weight (plf)	Area of steel (in.³)	depth (in.)	tw-web thickness (in.)	flange width (in.)	tf-flange thickness (in.)	Ix (in.⁴)	Zx (in.³)	Sx (in.³)	Zy (in.³)	Sy (in.³)
W27x146	146	43.1	27.38	0.605	13.97	0.975	5660	464	414	97.7	63.5
W27x129	129	37.8	27.63	0.61	10.01	1.1	4760	395	345	57.6	36.8
W27x114	114	33.5	27.29	0.57	10.07	0.93	4080	343	299	49.3	31.5
W27x102	102	30	27.09	0.515	10.01	0.83	3620	305	267	43.4	27.8
W27x94	94	27.7	26.92	0.49	9.99	0.745	3270	278	243	38.8	24.8
W27x84	84	24.8	26.71	0.46	9.96	0.64	2850	244	213	33.2	21.2
W24x192	192	56.3	25.47	0.81	12.95	1.46	6260	559	491	126	81.8
W24x176	176	51.7	25.24	0.75	12.89	1.34	5680	511	450	115	74.3
W24x162	162	47.7	25	0.705	12.96	1.22	5170	468	414	105	68.4
W24x146	146	43	24.74	0.65	12.9	1.09	4580	418	371	93.2	60.5
W24x131	131	38.5	24.48	0.605	12.86	0.96	4020	370	329	81.5	53
W24x117	117	34.4	24.26	0.55	12.8	0.85	3540	327	291	71.4	46.5
W24x104	104	30.6	24.06	0.5	12.75	0.75	3100	289	258	62.4	40.7
W24x103	103	30.3	24.53	0.55	9	0.98	3000	280	245	41.5	26.5
W24x94	94	27.7	24.31	0.515	9.07	0.875	2700	254	222	37.5	24
W24x84	84	24.7	24.1	0.47	9.02	0.77	2370	224	196	32.6	20.9
W24x76	76	22.4	23.92	0.44	8.99	0.68	2100	200	176	28.6	18.4
W24x68	68	20.1	23.73	0.415	8.97	0.585	1830	177	154	24.5	15.7
W24x62	62	18.3	23.74	0.43	7.04	0.59	1560	154	132	15.8	9.8
W24x55	55	16.3	23.57	0.395	7.01	0.505	1360	135	115	13.4	8.3
W21x147	147	43.2	22.06	0.72	12.51	1.15	3630	373	329	92.6	60.1
W21x132	132	38.8	21.83	0.65	12.44	1.03	3220	333	295	82.3	53.5
W21x122	122	35.9	21.68	0.6	12.39	0.96	2960	307	273	75.6	49.2
W21x111	111	32.7	21.51	0.55	12.34	0.875	2670	279	249	68.2	44.5
W21x101	101	29.8	21.36	0.5	12.29	0.8	2420	253	227	61.7	40.3

FIGURE 18.2 Wide flange tables. *(Courtesy of Seaport Steel, Inc. Seattle, WA)*

Designation	weight (plf)	Area of steel (in.^{3})	depth (in.)	tw-web thickness (in.)	flange width (in.)	tf-flange thickness (in.)	Ix (in.^{4})	Zx (in.^{3})	Sx (in.^{3})	Zy (in.^{3})	Sy (in.^{3})
W21x93	93	27.3	21.62	0.58	8.42	0.93	2070	221	192	34.7	22.1
W21x83	83	24.3	21.43	0.515	8.36	0.835	1830	196	171	30.5	19.5
W21x73	73	21.5	21.24	0.455	8.29	0.74	1600	172	151	26.6	17
W21x68	68	20	21.13	0.43	8.27	0.685	1480	160	140	24.4	15.7
W21x62	62	18.3	20.99	0.4	8.24	0.615	1330	144	127	21.7	14
W21x55	55	16.2	20.8	0.375	8.22	0.522	1140	126	110	18.4	11.8
W21x48	48	14.1	20.6	0.35	8.14	0.43	959	107	93	14.9	9.52
W21x57	57	16.7	21.06	0.405	6.56	0.65	1170	129	111	14.8	9.35
W21x50	50	14.7	20.83	0.38	6.53	0.535	984	110	94.5	12.2	7.64
W21x44	44	13	20.66	0.35	6.5	0.45	843	95.4	81.6	10.2	6.37
W18x119	119	35.1	18.97	0.655	11.27	1.06	2190	262	231	69.1	44.9
W18x106	106	31.1	18.73	0.59	11.2	0.94	1910	230	204	60.5	39.4
W18x97	97	28.5	18.59	0.535	11.15	0.87	1750	211	188	55.3	36.1
W18x86	86	25.3	18.39	0.48	11.09	0.77	1530	186	166	48.4	31.6
W18x76	76	22.3	18.21	0.425	11.04	0.68	1330	163	146	42.2	27.6
W18x71	71	20.8	18.47	0.495	7.64	0.81	1170	146	127	24.7	15.8
W18x65	65	19.1	18.35	0.45	7.59	0.75	1070	133	117	22.5	14.4
W18x60	60	17.6	18.24	0.415	7.56	0.695	984	123	108	20.6	13.3
W18x55	55	16.2	18.11	0.39	7.53	0.63	890	112	98.3	18.5	11.9
W18x50	50	14.7	17.99	0.355	7.49	0.57	800	101	88.9	16.6	10.7
W18x46	46	13.5	18.06	0.36	6.06	0.605	712	90.7	78.8	11.7	7.43
W18x40	40	11.8	17.9	0.315	6.02	0.525	612	78.4	68.4	9.95	6.35
W18x35	35	10.3	17.7	0.3	6	0.425	510	66.5	57.6	8.06	5.12
W16x100	100	29.7	16.97	0.585	10.43	0.985	1500	200	177	55	35.7
W16x89	89	26.4	16.75	0.525	10.37	0.875	1310	177	157	48.2	31.4
W16x77	77	22.9	16.52	0.455	10.29	0.76	1120	152	136	41.2	26.9
W16x67	67	20	16.33	0.395	10.24	0.665	970	132	119	35.6	23.2
W16x57	57	16.8	16.43	0.43	7.12	0.715	758	105	92.2	18.9	12.1
W16x50	50	14.7	16.26	0.38	7.07	0.63	659	92	81	16.3	10.5
W16x45	45	13.3	16.13	0.345	7.04	0.565	586	82.3	72.7	14.5	9.34
W16x40	40	11.8	16.01	0.305	6.99	0.505	518	73	64.7	12.7	8.25
W16x36	36	10.6	15.86	0.295	6.98	0.43	448	64	56.5	10.8	7
W14x193	193	56.8	15.48	0.89	15.71	1.44	2400	355	310	180	119
W14x176	176	51.8	15.22	0.83	15.65	1.31	2140	320	281	163	107
W14x159	159	46.7	14.98	0.745	15.57	1.19	1900	287	254	146	96.2
W14x145	145	42.7	14.78	0.68	15.5	1.09	1710	260	232	133	87.3
W14x132	132	38.8	14.66	0.645	14.73	1.03	1530	234	209	113	74.5
W14x120	120	35.3	14.48	0.59	14.67	0.94	1380	212	190	102	67.5
W14x109	109	32	14.32	0.525	14.61	0.86	1240	192	173	92.7	61.2
W14x99	99	29.1	14.16	0.485	14.66	0.78	1110	173	157	83.6	55.2
W14x90	90	26.5	14.02	0.44	14.52	0.71	999	157	143	75.6	49.9

FIGURE 18.2 *(continued)* Wide flange tables. *(Courtesy of Seaport Steel, Inc. Seattle, WA)*

504 EARTH RETENTION SYSTEMS

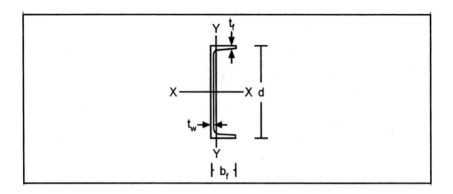

CHANNEL TABLES

Designation	weight steel (plf)	Area of (in.^{2})	depth 'd' (in.)	tw-web thickness (in.)	flange width-bf (in.)	tf-flange thickness (in.)	Ix (in.^{4})	Zx (in.^{3})	Sx (in.^{3})	Zy (in.^{3})	Sy (in.^{3})
C15x50	50	14.7	15	0.716	3.72	0.65	404	68.5	53.8	8.14	3.77
C15x40	40	11.8	15	0.52	3.52	0.65	348	57.5	46.5	6.84	3.34
C15x33.9	33.9	9.95	15	0.4	3.4	0.65	315	50.8	42	6.19	3.09
C12x30	30	8.81	12	0.51	3.17	0.501	162	33.8	27	4.32	2.05
C12x25	25	7.34	12	0.387	3.05	0.501	144	29.4	24	3.82	1.87
C12x20.7	20.7	6.08	12	0.282	2.94	0.501	129	25.6	21.5	3.47	1.72
C10x30	30	8.81	10	0.673	3.03	0.436	103	26.7	20.7	3.78	1.65
C10x25	25	7.34	10	0.526	2.89	0.436	91.1	23.1	18.2	3.18	1.47
C10x20	20	5.87	10	0.379	2.74	0.436	78.9	19.4	15.8	2.7	1.31
C10x15.3	15.3	4.48	10	0.24	2.6	0.436	67.3	15.9	13.5	2.34	1.15
MC18x58	58	17.1	18	0.7	4.2	0.625	675	95.4	75	10.7	5.28
MC18x51.9	51.9	15.3	18	0.6	4.1	0.625	627	87.3	69.6	9.86	5.02
MC18x45.8	45.8	13.5	18	0.5	4	0.625	578	79.2	64.2	9.14	4.77
MC18x42.7	42.7	12.6	18	0.45	3.95	0.625	554	75.1	61.5	8.82	4.64
MC13x50	50	14.7	13	0.787	4.41	0.61	314	60.8	48.3	10.2	4.77
MC13x40	40	11.8	13	0.56	4.19	0.61	273	51.2	41.9	8.66	4.24
MC13x35	35	10.3	13	0.447	4.07	0.61	252	46.5	38.8	8.04	3.97
MC13x31.8	31.8	9.35	13	0.375	4	0.61	239	43.4	36.7	7.69	3.79
MC12x50	50	14.7	12	0.835	4.14	0.7	269	56.5	44.9	10.9	5.64
MC12x45	45	13.2	12	0.712	4.01	0.7	252	52	41.9	10.1	5.31
MC12x40	40	11.8	12	0.59	3.89	0.7	234	47.7	39	9.31	4.98
MC12x35	35	10.3	12	0.467	3.77	0.7	216	43.2	36.1	8.63	4.65
MC12x31	31	9.12	12	0.37	3.67	0.7	202	39.7	33.7	8.15	4.37
MC10x41.1	41.1	12.1	10	0.796	4.32	0.575	157	39.3	31.5	9.49	4.85
MC10x33.6	33.6	9.87	10	0.575	4.1	0.575	139	33.7	27.8	8.28	4.35
MC10x28.5	28.5	8.37	10	0.425	3.95	0.575	126	30	25.3	7.59	3.99
MC10x25	25	7.35	10	0.38	3.41	0.575	110	26.2	22	5.65	2.96
MC10x22	22	6.45	10	0.29	3.32	0.575	102	23.9	20.5	5.29	2.75

FIGURE 18.3 Channel tables. *(Courtesy of Seaport Steel, Inc. Seattle, WA)*

Position Paper by the International Association of Drilled Shaft Contractors (ADSC) on Technical Aspects of Proposed AASHTO LRFD Specifications for Retaining Walls

ANCHOR/SOIL TYPE (GROUT PRESSURE)	SOIL STIFFNESS OR UNCONFINED COMPRESSIVE STRENGTH (TSF) OR SPT RESISTANCE {BLOWS/FT}	PRESUMPTIVE ULTIMATE BOND STRESS (KSF) (note 1,2)
COHESIVE SOILS		
Gravity Grouted anchors (<50 psi) (note 3)		
Silt-Clay Mixtures	Stiff to Very Stiff (1.0 - 4.0)	.7 - 1.5
Pressure Grouted Anchors (50 psi - 400 psi)		
High Plasticity Clay (note 4)	Stiff (1.0 - 2.5)	.7 - 3
High Plasticity Clay (note 4)	Very Stiff (2.5 - 4.0)	1.5 - 4
Med. Plasticity Clay	Stiff (1.0 - 2.5)	.7 - 3
Med. Plasticity Clay	Very Stiff (2.5 - 4.0)	1.5 - 4
Med. Plasticity Sandy Silt	Very Stiff (2.5 - 4.0)	1.5 - 8
COHESIONLESS SOILS		
Gravity Grouted anchors (<50 psi) (note 3)		
Sand or Sand-Gravel Mixture	Med. Dense to Dense {11-50}	1.5 - 3
Pressure Grouted Anchors (50 psi - 400 psi)		
Fine to Medium Sand	Med. Dense to Dense {11-50}	1.5 - 5
Medium to Coarse Sand w/Gravel	Med. Dense {11-30}	2.5 - 8
Medium to Coarse Sand w/Gravel	Dense to Very Dense {30-50+}	2.5 - 8
Silty Sands	---	1.5 - 5
Sandy Gravel	Med. Dense to Dense {11-40}	2.5 - 8
Sandy Gravel	Dense to Very Dense {40-50+}	2.5 - 8
Glacial Till	Dense {31-50}	2 - 7
ROCK		
Granite or Basalt		30 - 90
Dolomitic Limestone		20 - 45
Soft Limestone		15 - 30
Slates and Hard Shales		10 - 30
Sandstones		10 - 40
Weathered Sandstones		10 - 35
Soft Shales		4 - 12

TABLE 1

NOTES:

1. Values are approximate and should be used with engineering judgement. Consultation with an engineer on the staff of a specialty contractor with experience in the particular soil conditions at the site is recommended.

2. The value to be used for design should be the responsibility of the specialty contractor so as to allow for the use of proprietary and/or patented installation methods, etc. The actual capacity of each tieback must be verified with appropriate testing per AASHTO/AGC/ARTBA Joint Task Force 27 Specification for Tiebacks.

3. Gravity grouting (i.e. without pressure) not generally recommended for soil tiebacks.

4. High plasticity clays may be creep susceptible. See recommendations in FHWA/RD-82/047 "Tiebacks"

FIGURE 18.4 Tieback adhesion. *(Courtesy of ADSC-The International Association of Foundation Drilling. Dallas, TX)*

Diameter Inches	Surface Area Ft²/Ft	Volume Ft³/Ft	Diameter mm	Surface Area M²/M	Volume M³/M
1	0.262	0.0055	25	0.079	0.00049
2	0.524	0.0218	50	0.157	0.00196
4	1.047	0.0873	100	0.314	0.00785
5	1.309	0.1364	126	0.396	0.01247
6	1.571	0.1964	152	0.478	0.01815
8	2.094	0.3491	203	0.638	0.03237
10	2.618	0.5454	254	0.798	0.05067
12	3.142	0.7854	305	0.958	0.0730
16	4.189	1.3963	406	1.275	0.12946
18	4.712	1.722	457	1.436	0.16403

FIGURE 18.5 Grouting Volumes—Neat Quantities

Diameter Inches	Surface Area Ft²/Ft	End Area Ft²	Volume yd³/ft	Diameter mm	Surface Area M³/M	End Area M²	Volume M³/M
12	3.142	0.785	0.0291	305	0.958	0.0731	0.0731
16	4.189	1.396	0.0517	406	1.275	0.1295	0.129
18	4.712	1.767	0.0655	457	1.436	0.1640	0.164
24	6.283	3.142	0.116	610	1.916	0.2922	0.292
28	7.330	4.276	0.158	711	2.234	0.3970	0.397
30	7.854	4.909	0.182	762	2.394	0.4560	0.456
36	9.425	7.069	0.262	915	2.875	0.6576	0.658
42	10.996	9.621	0.356	1066	3.349	0.8925	0.892
48	12.566	12.566	0.465	1220	3.833	1.1690	1.169
54	14.137	15.904	0.589	1373	4.313	1.4806	1.481

FIGURE 18.6 Drilled Shaft Dimension—Neat Quantities

SOIL NAIL INSTALLATION FORM

Structure __Retaining Wall #10__ Nail Number __Row 1, #5__ Date __7/20/92__

Inspector __J. Smith__ Contract __SN-3223__ Location __SR5 Main Blvd__ Station __20+00__

Method __Open Hole__ Rig Type __Klem Advancing 6" Dia. Solid Stem Auger__

Hole Diameter __6 in.__ Inclination __15°__ Tolerance Deviation __None__

Remarks __0'-4' = SILTY SAND (Fill)__

__4'-20' = SANDY SILT (Native); No Groundwater__

NAIL

Bar Diameter __1 in.__ Bar Length __18 ft.__ Total Steel Area __0.85 sq. in.__

Bond Length __15 ft.__ Unbonded Length __3 ft.__ Design Load __24 Kips__

Remarks __Epoxy coating, centralizers, and installation__

__procedures conform with specifications__

GROUTING

Cu. Ft./Stroke __0.13__ Start Time __1:07 pm__ Finish Time __1:10 pm__

Pump Pressure __75 - 100 psi__

Pumped Volume __28 strokes__ x __0.13 cu. ft./stroke__ = __3.64 cu. ft.__

Hole Volume __15 ft.__ x __0.1964 sq. ft.__ = __2.95 cu. ft.__

Grout Take (Ratio Pump Vol./Hole Vol.) __1.23__

Remarks __Grout placed by Tremie pipe in 15 ft. bonded__

__length for proof test; no grouting problems__

UNBONDED ZONE BACKFILL

Date __7/24/92__ Placement Method __Tremie Pipe__

Start Time __8:00 am__ Finish Time __8:05 am__ Estimated Grout Volume __0.6 cu. ft.__

Remarks __Grout returns via vent tube - adequate grout backfill__

Soil Nail Accepted _____ Date __7/24/92__

Remarks __Soil nail installation, grouting, and__

__proof test results conform with specifications/drawings__

FIGURE 18.7 Sample forms used for reporting tiebacks or soil nails. (*Courtesy of Federal Highway Administration. Washington, DC*)

SOIL NAIL INSTALLATION SUMMARY

Project: _SR14-WALL #10_   Project No.: _SR14 V10_   Field Inspector: _

Soil Nail Number	Date	Diameter/ Method/ Length	Rig Type	Inclin- ation	Soils	Tendon Dia./Grade/ Length
R1-1	2-2-93	6"/OH/20'	HSA	15°	FILL/SILT	#8/60/21'
R1-2	2-2-93	6"/OH/20'	HSA	15°	FILL/SILT	#8/60/21'
R1-3	2-2-93	6"/OH/20'	HSA	15°	FILL/SILT	#8/60/21'
R1-4	2-2-93	6"/OH/20'	HSA	15°	SILT/SILT	#8/60/21'
R1-5	2-2-93	6"/OH/20'	HSA	15°	FILL/SILT	#8/60/21'
R1-6	2-2-93	6"/OH/20'	HSA	15°	FILL/SILT	#8/60/21'
R1-7	2-2-93	6"/OH/20'	HSA	15°	FILL/SILT	#8/60/21'
R1-8	2-2-93	6"/OH/20'	HSA	15°	FILL/SILT	#8/60/21'
R1-9	2-2-93	6"/OH/20'	HSA	15°	FILL/SILT	#8/60/21'
R1-10	2-2-93	6"/OH/20'	HSA	15°	FILL/SILT	#8/60/21'
R1-11	2-2-93	6"/OH/20'	HSA	15°	FILL/SILT	#8/60/21'
R2-1	2-14-93	6"/OH/22'	HSA	15°	SANDY SILT	#8/60/23'
R2-2	2-14-93	6"/OH/22'	HSA	15°	SANDY SILT	#8/60/23'
R2-3	2-14-93	6"/OH/22'	HSA	15°	SANDY SILT	#8/60/23'
R2-4	2-14-93	6"/OH/22'	HSA	15°	SANDY SILT	#8/60/23'
R2-5	2-14-93	6"/OH/22'	HSA	15°	SANDY SILT	#8/60/23'
R2-6	2-14-93	6"/OH/22'	HSA	15°	SANDY SILT	#8/60/23'
R2-7	2-14-93	6"/OH/22'	HSA	15°	SANDY SILT/SILTY CLAY	#8/60/23'
R2-8	2-14-93	6"/OH/22'	HSA	15°	SANDY SILT/SILTY CLAY	#8/60/23'
R2-9	2-14-93	6"/OH/22'	HSA	15°	SANDY SILT/SILTY CLAY	#8/60/23'
R2-10	2-14-93	6"/OH/22'	HSA	15°	SANDY SILT/SILTY CLAY	#8/60/23'
R2-11	2-14-93	6"/OH/22'	HSA	15°	SANDY SILT/SILTY CLAY	#8/60/23'

NOTES: HSA = HOLLOW STEM AUGER
          OH = OPEN HOLE

FIGURE 18.7 *(continued)* Sample forms used for reporting tiebacks or soil nails. *(Courtesy of Federal Highway Administration. Washington, DC)*

Adjustment Factors

Repetitive Member Factor, C_r

Bending design values, F_b, for dimension lumber 2" to 4" thick shall be multiplied by the repetitive member factor, $C_r = 1.15$, when such members are used as joists, truss chords, rafters, studs, planks, decking or similar members which are in contact or spaced not more than 24" on centers, are not less than 3 in number and are joined by floor, roof or other load distributing elements adequate to support the design load.

Wet Service Factor, C_M

When dimension lumber is used where moisture content will exceed 19% for an extended time period, design values shall be multiplied by the appropriate wet service factors from the following table:

Wet Service Factors, C_M

F_b	F_t	F_v	$F_{c\perp}$	F_c	E
0.85*	1.0	0.97	0.67	0.8**	0.9

* when $(F_b)(C_F) \leq 1150$ psi, $C_M = 1.0$
** when $(F_c)(C_F) \leq 750$ psi, $C_M = 1.0$

Flat Use Factor, C_{fu}

Bending design values adjusted by size factors are based on edgewise use (load applied to narrow face). When dimension lumber is used flatwise (load applied to wide face), the bending design value, F_b, shall also be multiplied by the following flat use factors:

Flat Use Factors, C_{fu}

Width (depth)	Thickness (breadth)	
	2" & 3"	4"
2" & 3"	1.0	—
4"	1.1	1.0
5"	1.1	1.05
6"	1.15	1.05
8"	1.15	1.05
10" & wider	1.2	1.1

NOTE

To facilitate the use of Table 4A, shading has been employed to distinguish design values based on a 4" nominal width (Construction, Standard and Utility grades) or a 6" nominal width (Stud grade) from design values based on a 12" nominal width (Select Structural, No.1 & Btr, No.1, No.2 and No.3 grades).

Size Factor, C_F

Tabulated bending, tension, and compression parallel to grain design values for dimension lumber 2" to 4" thick shall be multiplied by the following size factors:

Size Factors, C_F

Grades	Width (depth)	F_b		F_t	F_c
		Thickness (breadth)			
		2" & 3"	4"		
Select Structural, No. 1 & Btr, No. 1, No. 2, No. 3	2", 3" & 4"	1.5	1.5	1.5	1.15
	5"	1.4	1.4	1.4	1.1
	6"	1.3	1.3	1.3	1.1
	8"	1.2	1.3	1.2	1.05
	10"	1.1	1.2	1.1	1.0
	12"	1.0	1.1	1.0	1.0
	14" & wider	0.9	1.0	0.9	0.9
Stud	2", 3" & 4"	1.1	1.1	1.1	1.05
	5" & 6"	1.0	1.0	1.0	1.0
	8" & wider	Use No. 3 Grade tabulated design values and size factors			
Construction & Standard	2", 3" & 4"	1.0	1.0	1.0	1.0
Utility	4"	1.0	1.0	1.0	1.0
	2" & 3"	0.4	—	0.4	0.6

FIGURE 18.8 Timber design specifications. *(Courtesy of American Forest and Paper Association. Washington, DC)*

Shear Stress Factor, C_H

Tabulated shear design values parallel to grain have been reduced to allow for the occurrence of splits, checks and shakes. Tabulated shear design values parallel to grain, F_v, shall be permitted to be multiplied by the shear stress factors specified in the following table when length of split, or size of check or shake is known and no increase in them is anticipated. When shear stress factors are used for Redwood, a tabulated design value of $F_v = 80$ psi shall be assigned for all grades of Redwood dimension lumber. Shear stress factors shall be permitted to be linearly interpolated.

Shear Stress Factors, C_H

Length of split on wide face of 2" (nominal) lumber	C_H	Length of split on wide face of 3" (nominal) and thicker lumber	C_H	Size of shake* in 2" (nominal) and thicker lumber	C_H
no split	2.00	no split	2.00	no shake	2.00
1/2 × wide face	1.67	1/2 × narrow face	1.67	1/6 × narrow face	1.67
3/4 × wide face	1.50	3/4 × narrow face	1.50	1/4 narrow face	1.50
1 × wide face	1.33	1 × narrow face	1.33	1/3 × narrow face	1.33
1-1/2 × wide face or more	1.00	1-1/2 × narrow face or more	1.00	1/2 × narrow face or more	1.00

*Shake is measured at the end between lines enclosing the shake and perpendicular to the loaded face.

Table 4A Base Design Values for Visually Graded Dimension Lumber
(All species except Southern Pine — see Table 4B) (Tabulated design values are for normal load duration and dry service conditions. See NDS 2.3 for a comprehensive description of design value adjustment factors.)

USE WITH TABLE 4A ADJUSTMENT FACTORS

Design values in pounds per square inch (psi)

Species and commercial grade	Size classification	Bending F_b	Tension parallel to grain F_t	Shear parallel to grain F_v	Compression perpendicular to grain $F_{c\perp}$	Compression parallel to grain F_c	Modulus of Elasticity E	Grading Rules Agency
ASPEN								
Select Structural		875	500	60	265	725	1,100,000	
No.1	2"-4" thick	625	375	60	265	600	1,100,000	
No.2		600	350	60	265	450	1,000,000	NELMA
No.3	2"& wider	350	200	60	265	275	900,000	NSLB
Stud		475	275	60	265	300	900,000	WWPA
Construction	2"-4" thick	700	400	60	265	625	900,000	
Standard		375	225	60	265	475	900,000	
Utility	2"-4" wide	175	100	60	265	300	800,000	
BEECH-BIRCH-HICKORY								
Select Structural		1450	850	100	715	1200	1,700,000	
No.1	2"-4" thick	1050	600	100	715	950	1,600,000	
No.2		1000	600	100	715	750	1,500,000	
No.3	2"& wider	575	350	100	715	425	1,300,000	NELMA
Stud		775	450	100	715	475	1,300,000	
Construction	2"-4" thick	1150	675	100	715	1000	1,400,000	
Standard		650	375	100	715	775	1,300,000	
Utility	2"-4" wide	300	175	100	715	500	1,200,000	
COTTONWOOD								
Select Structural		875	525	65	320	775	1,200,000	
No.1	2"-4" thick	625	375	65	320	625	1,200,000	
No.2		625	350	65	320	475	1,100,000	
No.3	2"& wider	350	200	65	320	275	1,000,000	NSLB
Stud		475	275	65	320	300	1,000,000	
Construction	2"-4" thick	700	400	65	320	650	1,000,000	
Standard		400	225	65	320	500	900,000	
Utility	2"-4" wide	175	100	65	320	325	900,000	

FIGURE 18.8 *(continued)* Timber design specifications. *(Courtesy of American Forest and Paper Association. Washington, DC)*

Base Design Values for Visually Graded Dimension Lumber (Cont.)

(All species except Southern Pine — see Table 4B) (Tabulated design values are for normal load duration and dry service conditions. See NDS 2.3 for a comprehensive description of design value adjustment factors.)

USE WITH TABLE 4A ADJUSTMENT FACTORS

Species and commercial grade	Size classification	Bending F_b	Tension parallel to grain F_t	Shear parallel to grain F_v	Compression perpendicular to grain $F_{c\perp}$	Compression parallel to grain F_c	Modulus of Elasticity E	Grading Rules Agency
DOUGLAS FIR-LARCH								
Select Structural		1500	1000	95	625	1700	1,900,000	
No.1 & Btr	2"-4" thick	1200	800	95	625	1550	1,800,000	
No.1		1000	675	95	625	1500	1,700,000	
No.2	2" & wider	900	575	95	625	1350	1,600,000	WCLIB
No.3		525	325	95	625	775	1,400,000	WWPA
Stud		700	450	95	625	850	1,400,000	
Construction	2"-4" thick	1000	650	95	625	1650	1,500,000	
Standard		575	375	95	625	1400	1,400,000	
Utility	2"-4" wide	275	175	95	625	900	1,300,000	
DOUGLAS FIR-LARCH (NORTH)								
Select Structural	2"-4" thick	1350	825	95	625	1900	1,900,000	
No.1/No.2		850	500	95	625	1400	1,600,000	
No.3	2" & wider	475	300	95	625	825	1,400,000	NLGA
Stud		650	400	95	625	900	1,400,000	
Construction	2"-4" thick	950	575	95	625	1800	1,500,000	
Standard		525	325	95	625	1450	1,400,000	
Utility	2"-4" wide	250	150	95	625	950	1,300,000	
DOUGLAS FIR-SOUTH								
Select Structural		1350	900	90	520	1600	1,400,000	
No.1	2"-4" thick	925	600	90	520	1450	1,300,000	
No.2		850	525	90	520	1350	1,200,000	
No.3	2" & wider	500	300	90	520	775	1,100,000	WWPA
Stud		675	425	90	520	850	1,100,000	
Construction	2"-4" thick	975	600	90	520	1650	1,200,000	
Standard		550	350	90	520	1400	1,100,000	
Utility	2"-4" wide	250	150	90	520	900	1,000,000	
EASTERN HEMLOCK-BALSAM FIR								
Select Structural		1250	575	70	335	1200	1,200,000	
No.1	2"-4" thick	775	350	70	335	1000	1,100,000	
No.2		575	275	70	335	825	1,100,000	
No.3	2" & wider	350	150	70	335	475	900,000	NELMA
Stud		450	200	70	335	525	900,000	NSLB
Construction	2"-4" thick	675	300	70	335	1050	1,000,000	
Standard		375	175	70	335	850	900,000	
Utility	2"-4" wide	175	75	70	335	550	800,000	
EASTERN HEMLOCK-TAMARACK								
Select Structural		1250	575	85	555	1200	1,200,000	
No.1	2"-4" thick	775	350	85	555	1000	1,100,000	
No.2		575	275	85	555	825	1,100,000	
No.3	2" & wider	350	150	85	555	475	900,000	NELMA
Stud		450	200	85	555	525	900,000	NSLB
Construction	2"-4" thick	675	300	85	555	1050	1,000,000	
Standard		375	175	85	555	850	900,000	
Utility	2"-4" wide	175	75	85	555	550	800,000	
EASTERN SOFTWOODS								
Select Structural		1250	575	70	335	1200	1,200,000	
No.1	2"-4" thick	775	350	70	335	1000	1,100,000	
No.2		575	275	70	335	825	1,100,000	
No.3	2" & wider	350	150	70	335	475	900,000	NELMA
Stud		450	200	70	335	525	900,000	NSLB
Construction	2"-4" thick	675	300	70	335	1050	1,000,000	
Standard		375	175	70	335	850	900,000	
Utility	2"-4" wide	175	75	70	335	550	800,000	

FIGURE 18.8 *(continued)* Timber design specifications. *(Courtesy of American Forest and Paper Association. Washington, DC)*

Base Design Values for Visually Graded Dimension Lumber (Cont.)
(All species except Southern Pine — see Table 4B) (Tabulated design values are for normal load duration and dry service conditions. See NDS 2.3 for a comprehensive description of design value adjustment factors.)

USE WITH TABLE 4A ADJUSTMENT FACTORS

			Design values in pounds per square inch (psi)					
Species and commercial grade	Size classification	Bending F_b	Tension parallel to grain F_t	Shear parallel to grain F_v	Compression perpendicular to grain $F_{c\perp}$	Compression parallel to grain F_c	Modulus of Elasticity E	Grading Rules Agency
EASTERN WHITE PINE								
Select Structural		1250	575	70	350	1200	1,200,000	
No.1	2"-4" thick	775	350	70	350	1000	1,100,000	
No.2		575	275	70	350	825	1,100,000	
No.3	2"& wider	350	150	70	350	475	900,000	NELMA
Stud		450	200	70	350	525	900,000	NSLB
Construction	2"-4" thick	675	300	70	350	1050	1,000,000	
Standard		375	175	70	350	850	900,000	
Utility	2"-4" wide	175	75	70	350	550	800,000	
HEM-FIR								
Select Structural		1400	925	75	405	1500	1,600,000	
No.1 & Btr	2"-4" thick	1100	725	75	405	1350	1,500,000	
No.1		975	625	75	405	1350	1,500,000	
No.2	2"& wider	850	525	75	405	1300	1,300,000	WCLIB
No.3		500	300	75	405	725	1,200,000	WWPA
Stud		675	400	75	405	800	1,200,000	
Construction	2"-4" thick	975	600	75	405	1550	1,300,000	
Standard		550	325	75	405	1300	1,200,000	
Utility	2"-4" wide	250	150	75	405	850	1,100,000	
HEM-FIR (NORTH)								
Select Structural	2"-4" thick	1300	775	75	370	1700	1,700,000	
No.1/No.2		1000	575	75	370	1450	1,600,000	
No.3	2"& wider	575	325	75	370	850	1,400,000	NLGA
Stud		775	450	75	370	925	1,400,000	
Construction	2"-4" thick	1150	650	75	370	1750	1,500,000	
Standard		650	350	75	370	1500	1,400,000	
Utility	2"-4" wide	300	175	75	370	975	1,300,000	
MIXED MAPLE								
Select Structural		1000	600	100	620	875	1,300,000	
No.1	2"-4" thick	725	425	100	620	700	1,200,000	
No.2		700	425	100	620	550	1,100,000	
No.3	2"& wider	400	250	100	620	325	1,000,000	NELMA
Stud		550	325	100	620	350	1,000,000	
Construction	2"-4" thick	800	475	100	620	725	1,100,000	
Standard		450	275	100	620	575	1,000,000	
Utility	2"-4" wide	225	125	100	620	375	900,000	
MIXED OAK								
Select Structural		1150	675	85	800	1000	1,100,000	
No.1	2"-4" thick	825	500	85	800	825	1,000,000	
No.2		800	475	85	800	625	900,000	
No.3	2"& wider	475	275	85	800	375	800,000	NELMA
Stud		625	375	85	800	400	800,000	
Construction	2"-4" thick	925	550	85	800	850	900,000	
Standard		525	300	85	800	650	800,000	
Utility	2"-4" wide	250	150	85	800	425	800,000	
NORTHERN RED OAK								
Select Structural		1400	800	110	885	1150	1,400,000	
No.1	2"-4" thick	1000	575	110	885	925	1,400,000	
No.2		975	575	110	885	725	1,300,000	
No.3	2"& wider	550	325	110	885	425	1,200,000	NELMA
Stud		750	450	110	885	450	1,200,000	
Construction	2"-4" thick	1100	650	110	885	975	1,200,000	
Standard		625	350	110	885	750	1,100,000	
Utility	2"-4" wide	300	175	110	885	500	1,000,000	

FIGURE 18.8 *(continued)* Timber design specifications. *(Courtesy of American Forest and Paper Association. Washington, DC)*

Base Design Values for Visually Graded Dimension Lumber (Cont.)

(All species except Southern Pine — see Table 4B) (Tabulated design values are for normal load duration and dry service conditions. See NDS 2.3 for a comprehensive description of design value adjustment factors.)

USE WITH TABLE 4A ADJUSTMENT FACTORS

Species and commercial grade	Size classification	Bending F_b	Tension parallel to grain F_t	Shear parallel to grain F_v	Compression perpendicular to grain $F_{c\perp}$	Compression parallel to grain F_c	Modulus of Elasticity E	Grading Rules Agency
NORTHERN SPECIES								
Select Structural	2"-4" thick	1000	450	65	350	1100	1,100,000	
No.1/No.2		600	275	65	350	850	1,100,000	
No.3	2"& wider	350	150	65	350	500	1,000,000	NLGA
Stud		475	225	65	350	550	1,000,000	
Construction	2"-4" thick	700	300	65	350	1050	1,000,000	
Standard		400	175	65	350	875	900,000	
Utility	2"-4" wide	175	75	65	350	575	900,000	
NORTHERN WHITE CEDAR								
Select Structural		775	450	60	370	750	800,000	
No.1	2"-4" thick	575	325	60	370	600	700,000	
No.2		550	325	60	370	475	700,000	
No.3	2"& wider	325	175	60	370	275	600,000	NELMA
Stud		425	250	60	370	300	600,000	
Construction	2"-4" thick	625	375	60	370	625	700,000	
Standard		350	200	60	370	475	600,000	
Utility	2"-4" wide	175	100	60	370	325	600,000	
RED MAPLE								
Select Structural		1300	750	105	615	1100	1,700,000	
No.1	2"-4" thick	925	550	105	615	900	1,600,000	
No.2		900	525	105	615	700	1,500,000	
No.3	2"& wider	525	300	105	615	400	1,300,000	NELMA
Stud		700	425	105	615	450	1,300,000	
Construction	2"-4" thick	1050	600	105	615	925	1,400,000	
Standard		575	325	105	615	725	1,300,000	
Utility	2"-4" wide	275	150	105	615	475	1,200,000	
RED OAK								
Select Structural		1150	675	85	820	1000	1,400,000	
No.1	2"-4" thick	825	500	85	820	825	1,300,000	
No.2		800	475	85	820	625	1,200,000	
No.3	2"& wider	475	275	85	820	375	1,100,000	NELMA
Stud		625	375	85	820	400	1,100,000	
Construction	2"-4" thick	925	550	85	820	850	1,200,000	
Standard		525	300	85	820	650	1,100,000	
Utility	2"-4" wide	250	150	85	820	425	1,000,000	
REDWOOD								
Clear Structural		1750	1000	145	650	1850	1,400,000	
Select Structural		1350	800	80	650	1500	1,400,000	
Select Structural, open grain		1100	625	80	425	1100	1,100,000	
No.1	2"-4" thick	975	575	80	650	1200	1,300,000	
No.1, open grain		775	450	80	425	900	1,100,000	
No.2	2"& wider	925	525	80	650	950	1,200,000	
No.2, open grain		725	425	80	425	700	1,000,000	RIS
No.3		525	300	80	650	550	1,100,000	
No.3, open grain		425	250	80	425	400	900,000	
Stud		575	325	80	425	450	900,000	
Construction	2"-4" thick	825	475	80	425	925	900,000	
Standard		450	275	80	425	725	900,000	
Utility	2"-4" wide	225	125	80	425	475	800,000	
SPRUCE-PINE-FIR								
Select Structural	2"-4" thick	1250	700	70	425	1400	1,500,000	
No.1/No.2		875	450	70	425	1150	1,400,000	
No.3	2"& wider	500	250	70	425	650	1,200,000	NLGA
Stud		675	350	70	425	725	1,200,000	
Construction	2"-4" thick	1000	500	70	425	1400	1,300,000	
Standard		550	275	70	425	1150	1,200,000	
Utility	2"-4" wide	275	125	70	425	750	1,100,000	

FIGURE 18.8 *(continued)* Timber design specifications. *(Courtesy of American Forest and Paper Association. Washington, DC)*

Base Design Values for Visually Graded Dimension Lumber (Cont.)

(All species except Southern Pine — see Table 4B) (Tabulated design values are for normal load duration and dry service conditions. See NDS 2.3 for a comprehensive description of design value adjustment factors.)

USE WITH TABLE 4A ADJUSTMENT FACTORS

Species and commercial grade	Size classification	Bending F_b	Tension parallel to grain F_t	Shear parallel to grain F_v	Compression perpendicular to grain $F_{c\perp}$	Compression parallel to grain F_c	Modulus of Elasticity E	Grading Rules Agency
SPRUCE-PINE-FIR (SOUTH)								
Select Structural		1300	575	70	335	1200	1,300,000	
No.1	2"-4" thick	875	400	70	335	1050	1,200,000	
No.2		775	350	70	335	1000	1,100,000	NELMA
No.3	2"& wider	450	200	70	335	575	1,000,000	NSLB
Stud		600	275	70	335	625	1,000,000	WCLIB
Construction	2"-4" thick	875	400	70	335	1200	1,000,000	WWPA
Standard		500	225	70	335	1000	900,000	
Utility	2"-4" wide	225	100	70	335	675	900,000	
WESTERN CEDARS								
Select Structural		1000	600	75	425	1000	1,100,000	
No.1	2"-4" thick	725	425	75	425	825	1,000,000	
No.2		700	425	75	425	650	1,000,000	
No.3	2"& wider	400	250	75	425	375	900,000	WCLIB
Stud		550	325	75	425	400	900,000	WWPA
Construction	2"-4" thick	800	475	75	425	850	900,000	
Standard		450	275	75	425	650	800,000	
Utility	2"-4" wide	225	125	75	425	425	800,000	
WESTERN WOODS								
Select Structural		900	400	70	335	1050	1,200,000	
No.1	2"-4" thick	675	300	70	335	950	1,100,000	
No.2		675	300	70	335	900	1,000,000	
No.3	2"& wider	375	175	70	335	525	900,000	WCLIB
Stud		525	225	70	335	575	900,000	WWPA
Construction	2"-4" thick	775	350	70	335	1100	1,000,000	
Standard		425	200	70	335	925	900,000	
Utility	2"-4" wide	200	100	70	335	600	800,000	
WHITE OAK								
Select Structural		1200	700	110	800	1100	1,100,000	
No.1	2"-4" thick	875	500	110	800	900	1,000,000	
No.2		850	500	110	800	700	900,000	
No.3	2"& wider	475	275	110	800	400	800,000	NELMA
Stud		650	375	110	800	450	800,000	
Construction	2"-4" thick	950	550	110	800	925	900,000	
Standard		525	325	110	800	725	800,000	
Utility	2"-4" wide	250	150	110	800	475	800,000	
YELLOW POPLAR								
Select Structural		1000	575	75	420	900	1,500,000	
No.1	2"-4" thick	725	425	75	420	725	1,400,000	
No.2		700	400	75	420	575	1,300,000	
No.3	2"& wider	400	225	75	420	325	1,200,000	NSLB
Stud		550	325	75	420	350	1,200,000	
Construction	2"-4" thick	800	475	75	420	750	1,300,000	
Standard		450	250	75	420	575	1,100,000	
Utility	2"-4" wide	200	125	75	420	375	1,100,000	

1. **LUMBER DIMENSIONS.** Tabulated design values are applicable to lumber that will be used under dry conditions such as in most covered structures. For 2" to 4" thick lumber the DRY dressed sizes shall be used (see Table 1A) regardless of the moisture content at the time of manufacture or use. In calculating design values, the natural gain in strength and stiffness that occurs as lumber dries has been taken into consideration as well as the reduction in size that occurs when unseasoned lumber shrinks. The gain in load carrying capacity due to increased strength and stiffness resulting from drying more than offsets the design effect of size reductions due to shrinkage.
2. **STRESS-RATED BOARDS.** Stress-rated boards of nominal 1", 1¼" and 1½" thickness, 2" and wider, of most species, are permitted the design values shown for Select Structural, No.1 & Btr, No.1, No.2, No.3, Stud, Construction, Standard, Utility, Clear Heart Structural and Clear Structural grades as shown in the 2" to 4" thick categories herein, when graded in accordance with the stress-rated board provisions in the applicable grading rules. Information on stress-rated board grades applicable to the various species is available from the respective grading rules agencies. Information on additional design values may also be available from the respective grading agencies.

FIGURE 18.8 *(continued)* Timber design specifications. *(Courtesy of American Forest and Paper Association. Washington, DC)*

```
MIX ID : 7530 [    ]           CONCRETE MIX DESIGN                    12/18/01
                                    4000 PSI


CONTRACTOR          :  Condon Johnson
SOURCE OF CONCRETE  :  Plant #10 Renton & #12 Tukwila
CONSTRUCTION TYPE   :  Shotcrete


         WEIGHTS PER CUBIC YARD     (SATURATED, SURFACE-DRY)
                                                          YIELD, CU FT
Ash Grove Type I-II ASTM C 150, LB              705              3.59
ISG Resourses Fly Ash ASTM C 618 Class F, LB     30              0.21
Concrete Sand   Pit # A-464, LB                1890             11.65
AASHTO #8       Pit # A-464, LB                 910              5.34
WATER, LB (GAL-US)                              315 ( 37.7)      5.05
TOTAL AIR, %                                    5.0 +/- 1.5      1.36
                                                               ======
                                                    TOTAL       27.20

W.R. Grace Daravair  ASTM C 260, OZ-US           7.4

WATER/CEMENT RATIO, LBS/LB                       0.43
SLUMP, IN                                        2.50
CONCRETE UNIT WEIGHT, PCF                      141.5
```

FIGURE 18.9 Sample concrete mix designs. *(Courtesy of Stoneway Concrete. Renton, WA)*

MIX ID : 1510 [   ]   CONCRETE MIX DESIGN   12/18/01
                                  100 PSI

CONTRACTOR   : Condon Johnson
SOURCE OF CONCRETE : Plant #10 Renton & #12 Tukwila
CONSTRUCTION TYPE : Lean Mix
PLACEMENT   : Chute

WEIGHTS PER CUBIC YARD   (SATURATED, SURFACE-DRY)
                                             YIELD, CU FT
Ash Grove Type I-II, LB   145   0.74
Concrete Sand Class 2, LB   3051   18.80
WATER, LB (GAL-US)   433 (51.9)   6.94
TOTAL AIR, %   3.0   0.82
                                              ========
                                     TOTAL   27.30

WATER/CEMENT RATIO, LBS/LB   2.99
SLUMP, IN   6.00
CONCRETE UNIT WEIGHT, PCF   132.9

FIGURE 18.9 (*continued*) Sample concrete mix designs. (*Courtesy of Stoneway Concrete. Renton, WA*)

```
MIX ID : 751 [    ]        CONCRETE MIX DESIGN                    12/18/01
                              100 PSI

CONTRACTOR        : Condon Johnson
SOURCE OF CONCRETE: Plant #10 Renton & #12 Tukwila
CONSTRUCTION TYPE : Control Density Fill

         WEIGHTS PER CUBIC YARD    (SATURATED, SURFACE-DRY)
                                                    YIELD, CU FT
Ashgrove Type I-II ASTM C 150, LB              70         0.36
ISG Resources Fly Ash ASTM C 618 Class F, LB  300         2.14
CDF Sand, LB                                 2836        17.48
WATER, LB (GAL-US)                            400 ( 47.9) 6.41
TOTAL AIR, %                                    3.0       0.82
                                                      =======
                                            TOTAL        27.20

WATER/CEMENT RATIO, LBS/LB                     1.08
SLUMP, IN                                     10.00
CONCRETE UNIT WEIGHT, PCF                    132.6
```

FIGURE 18.9 (*continued*) Sample concrete mix designs. (*Courtesy of Stoneway Concrete. Renton, WA*)

CONCRETE MIX DESIGN

MIX ID : 9002 [   ]   4000 PSI   12/18/01

CONTRACTOR   : Condon Johnson
SOURCE OF CONCRETE : Plant #10 Renton & #12 Tukwila
CONSTRUCTION TYPE : Augercast Piles

```
        WEIGHTS PER CUBIC YARD      (SATURATED, SURFACE-DRY)
                                                     YIELD, CU FT
Ash Grove Type I-II   ASTM C 150, LB       846            4.30
Concrete Sand     ASTM C 33, LB           2498           15.40
WATER, LB  (GAL-US)                        417 ( 50.0)    6.68
TOTAL AIR, %                                 3.0          0.82
                                                      =========
                                              TOTAL      27.20

Master Builders 100-XR ASTM C 494 Type B, OZ   33.84

WATER/CEMENT RATIO, LBS/LB                 0.49
SLUMP, IN                                  8.00
CONCRETE UNIT WEIGHT, PCF                138.3
```

FIGURE 18.9 *(continued)* Sample concrete mix designs. *(Courtesy of Stoneway Concrete, Renton, WA)*

```
MIX ID : 5894 [    ]                CONCRETE MIX DESIGN                 12/18/01
                                        4000 PSI

        CONTRACTOR      :  Condon Johnson
        SOURCE OF CONCRETE : Plant #10 Renton & #12 Tukwila
        CONSTRUCTION TYPE  : Soldier Pile Structural Toe

            WEIGHTS PER CUBIC YARD     (SATURATED, SURFACE-DRY)
                                                          YIELD, CU FT
Ash Grove Type I-II ASTM C 150, LB              544              2.77
ISG Resources Fly Ash ASTM C 618 Class F, LB    181              1.29
Concrete Sand ASTM C 33, LB                    1196              7.37
3/8"     ASTM C 33 #8, LB                      1897             11.14
WATER, LB    (GAL-US)                           255  ( 30.6)     4.09
TOTAL AIR, %                                    2.0              0.54
                                                            ==========
                                                   TOTAL        27.20

W.R. Grace WRDA-64 ASTM C 494 Type A, OZ-US    36.25
Master Builders 100XR ASTM C 494 Type D, OZ-   21.75

WATER/CEMENT RATIO, LBS/LB                      0.35
SLUMP, IN                                       8.00
CONCRETE UNIT WEIGHT, PCF                     149.8
```

FIGURE 18.9 *(continued)* Sample concrete mix designs. *(Courtesy of Stoneway Concrete. Renton, WA)*

CHAPTER 19
BIBLIOGRAPHY

1. ADSC. 1995. Standards and Specifications for the Foundation Drilling Industry, ADSC.
2. ADSC Geo-Support Committee. 2000, ADSC. Mechanical Anchor Product Data Committee, ADSC.
3. Armour and Groneck. 2000. Micropile Design And Construction Guidelines, FHWA.
4. Carson. 1965. Foundation Construction, McGraw-Hill.
5. Cheney. 1984. Permanent Ground Anchors, FHWA.
6. Cross, Wilbur. 1985. 75 Years of Foundation Engineering, The Benjamin Co.
7. Goldberg, Jaworski, and Gordon. 1976. Lateral Support Systems and Underpinning Vol II, Design Fundamentals, FHWA.
8. Goldberg-Zoino and Associates. 1976. Lateral Support Systems and Underpinning Vol. III, Construction Methods, FHWA.
9. Golder Associates. 1994. Soil Nailing Field Inspectors Manual, FHWA.
10. Golder Associates. 1996. Manual for Design and Construction Monitoring of Soil Nail Walls, FHWA.
11. Greer and Gardner. 1986. Construction of Drilled Pier Foundations, John Wiley and Sons.
12. GSP #25. 1990. Design and Performance of Earth Retaining Structures, ASCE.
13. GSP #33. 1992. Excavation and Support for the Urban Infrastructure, ASCE.
14. GSP #50. Foundation Upgrading and Repair, ASCE.

15. GSP #74. 1997. Guidelines of Engineering Practice for Braced and Tied-Back Excavations, ASCE.
16. GSP #83. 1998. Design and Construction of Earth Retaining Structures, ASCE.
17. Havers and Stubbs. 1971. Handbook of Heavy Construction, McGraw-Hill.
18. NAVFAC. 1982. Foundations and Earth Structures Design Manual 7.2, Department of the Navy.
19. O'Neill and Reese. 1999. Drilled Shafts: Construction Procedures and Design Methods, ADSC.
20. Papers by Kerisel, Peck. 1985. Proceedings of the Eleventh International Conference on Soil Mechanics and Foundation Engineering, A. A. Balkema.
21. Peck, Hanson, & Thorburn. 1953. Foundation Engineering. John Wiley and Sons.
22. Post Tensioning Institute. 1996. Recommendations for Prestressed Rock and Soil Anchors, PTI.
23. Sabatini, Pass, and Bachus. 1999. Ground Anchors and Anchored Systems, FHWA.
24. Terzaghi and Peck. 1948. Soil Mechanics in Engineering Practice, John Wiley and Sons.
25. USACE #15. Design of Sheet Pile Walls, ASCE.
26. Winterkorn and Fang. 1975. Foundation Engineering Handbook, Van Nostrand Reinhold Co.
27. WSDOT. 2001. Special Provisions Soil Anchors, Washington State Department of Transportation.
28. Wu, T.H. 1966. Soil Mechanics, Allyn and Bacon Inc.
29. Xanthakos, 1979, Slurry Walls, McGraw-Hill.
30. Xanthakos. 1991. Ground Anchors and Anchored Structures, John Wiley and Sons.

GLOSSARY

Active zone That portion of soil behind the shoring wall which is deemed to be subject to possible movement.

Anchor An embedded tension element which provides lateral support to a shoring wall through frictional attachment to soil or rock. The term includes all portions of the installation including the grout and anchorage.

Anchor head The device which attaches a strand anchor to the retaining wall.

Anchor plate A plate which the anchor head or nut is seated on.

Anchorage Name of the total assembly at the face of the shoring wall which includes, in the case of strand anchors, the anchor head, plate, wedges and corrosion protection cap (if used). For bar anchors it includes the plate, nut and corrosion protection cap (if used).

Bar Type of anchor tendon utilizing the rod form of steel.

Boulders Discrete elements of rock which are larger in their largest dimension than 12 inches (305 mm).

CDF *see Controlled density fill.*

Clay A sedimentary material with grains less than 0.002 millimeters in diameter.

Cobbles discrete elements of rock which are smaller than boulders and larger than gravel.

Concrete A manufactured mixture of water, Portland cement, sand, and gravel. May include fly ash and additives.

Controlled density fill a manufactured mixture of sand, water, Portland cement and fly ash which is used as a backfill substance for trenches and soldier piles.

Corner brace A horizontal compression element which provides lateral support for a portion of one shoring wall by bracing against a portion of the shoring wall oriented at 90 degrees to the first wall.

Deadman An embedded element which when attached to an anchor tendon, will provide lateral support to the anchor through passive resistance.

Drain *see Drainage fabric.*

Drain board *see Drainage fabric.*

Drainage fabric A manufactured material which facilitates movement of ground water. Consists of a plastic core incorporating drainage paths protected by a filter fabric. Used to provide drainage behind shotcrete walls and between lagging and cast-in-place concrete.

Gravel a mixture of discrete rock elements which are larger than sand sizes and less than 3 inches (75 mm) in their largest dimension.

Grout A manufactured mixture of water and cement which may or may not include sand aggregate.

Hard rock concrete Concrete used in the toe of soldier piles.

Jack A hydraulic device used to tension tieback anchors or soil nails for purposes of testing or stressing.

Kip A unit of force equal to 1000 pounds.

Lagging Elements used to span between soldier piles to retain exposed soil. Usually of timber but may be of cast-in place concrete, precast concrete, steel or plastic.

Lean mix A low strength mixture of water, sand and Portland cement designed for application as backfill in situations where the material must be subsequently removed by hand.

Load cell A hydraulic or electric device which measures the load applied by the jack to an anchor or soil nail.

Nail *see Soil nail.*

Nut A machined threaded steel piece which attaches a bar anchor to a shoring wall.

Raker A sloping compression element which provides lateral support to a shoring wall by bracing against a footing or other structural elements within the excavation.

Raking shore *see Raker.*

Ram *see Jack.*

Rankine wedge In earth pressure theory, that portion of the soil behind a shoring wall which provides the driving force attempting to overturn the shoring wall.

Rock A massive, naturally formed aggregate of mineral matter. One of the primary elements of the earth's crust.

Rock anchor An anchor which derives its capacity through frictional resistance from within the rock mass into which it is grouted.

Roll chock A welded attachment used to prevent the rolling of a waler caused by eccentric loading.

Sand A loose sedimentary substance composed of rock elements whose largest dimension is no greater than 2.0 millimeters and smallest dimension is no less than 0.06 millimeters.

Sheath A plastic or HDPE tube which covers an anchor tendon or discrete elements of the tendon and prevents adhesion of grout to the tendon or tendon element.

Shotcrete A concrete which is spray-applied using compressed air as the driving force.

Silt A sedimentary material with discrete element sizes between those of clay and sand.

Soil A general term used to a describe a sedimentary material which may consist of clay, silt, sand or gravel or any combination of these elements.

Soil anchor An anchor which derives its capacity through frictional attachment to the soil mass into which it is grouted or otherwise attached.

Soil nail An element placed in a cut face, usually by drilled and grouted methods, which provides added stiffness and strength to the soil mass

Soldier pile A vertical element usually made of steel which forms the primary framing of a shoring wall.

Spacer A small device of plastic or steel which forms a template for the setting of soil nails, anchor tendons or their elements.

Strand Flexible cable made of twisted high tensile wire used in post tensioning and the creation of anchor tendons.

Stressing jack *see Jack*.

Structural concrete Concrete which has a compressive strength of greater than 3000 psi (20 Mpa).

Strut A horizontal compression element which provides lateral support by bracing a portion of one wall against another portion which is parallel to the first.

Tendon The elongated portion of a soil or rock anchor placed in a drilled hole composed of strand or bar together with sheathing, encapsulation and spacers.

Tiebacks General term used to describe soil anchors, rock anchors and deadman anchors.

Toe That portion of a shoring wall which is below the level of the excavation.

Trench A long narrow excavation usually excavated for purposes of placing a utility.

Trenchbox A manufactured moveable device placed in a trench for purposes of providing sidewall stability and protection of personnel.

Trumpet A tube of plastic or steel attached to the back of the anchor plate which, when filled with corrosion inhibiting materials, will provide corrosion protection for the transition zone of the tendon from the encapsulation to the anchor head.

Wale(s) *see Waler.*

Waler(s) Steel horizontal elements, functioning in bending, used in shoring applications to spread concentrated loads from struts, tiebacks, or corner braces to the shoring wall.

Washers Steel castings which seat a nut used to attach a bar to its anchor plate.

Wedges Machined steel pieces which attach strand elements to anchor heads.

INDEX

A

ABI, 347, 348
Active zone, 131, 287, 301, 14, 316
Active pressure, 66, 223, 262, 270-273, 275, 278, 304, 307, 308, 311, 312, 322, 496
ADSC, 5, 132, 505
Air hammers, 2, 345
Alignment load, 395, 399, 403
American Piledriving Equipment (APE), 346
Anchor, 4, 96, 132, 134-138, 140, 142, 143, 145-151, 156, 159-161, 170, 179, 219, 279, 315, 316, 393, 395, 399-402, 404, 405, 496, 505
Anchor head, 132, 136, 140, 142, 395
Anchor length, 132
Applied load, 95
Anchor plate, 315
Anchorage, 4, 95, 170, 314, 315, 324, 401
Anchored earth retention, 5
Apparent cohesion, 191, 261, 311, 322, 324
Apparent earth pressure, 199, 250, 267, 276-279, 495
Aquifer, 325
Arching, 27, 51, 206, 264, 265, 311, 319, 322
At-rest pressure, 223, 263, 279
Auger casting, 149, 357, 363, 515
Augered shafts, 3

B

Backhoe, 214
Bar, 132, 134, 137, 138, 140, 142, 143, 145, 170, 199, 280
Basal heave, 14, 27, 280, 308
Base instability, 27, 51, 52, 283, 291, 299, 300
Basal stability, 65, 280, 307, 308
Bayshore, 355, 364
Belled anchor, 4
Bentonite, 4
Berlin method, 2
Berm, 52, 57, 58, 100, 102, 112, 216, 218
Bodine hammer, 2
Boil, 299, 300
Bond length, 132, 142, 149, 496

Bond stress, 132, 134
Bond zone, 132, 138, 150, 294, 316
Boussinesq, 284
Bulkhead, 279
Buoyant weight, 264, 272

C

Caisson, 2, 5, 6, 80
Calweld, 349
Cantilever, 3, 66, 67, 72, 98, 100, 101, 175-177, 185, 206, 210, 211, 241, 268, 269, 273, 278, 279, 291, 292, 299, 303, 305, 307, 310, 315, 495
Carousel, 363, 368
Casing, 4, 85, 144, 149, 150
Caterpillar, 384
Chicago caisson, 2
Chimneying, 299, 336
Chuck, 363
Cofferdam, 5, 25, 27, 28
Cohesion, 191, 261, 269, 270, 439, 440, 505
Cold rolled, 26, 32
Colloidal mixer, 363, 373
Cone, 3
Cone penetration test (CPT), 250, 260, 261
Continuous flight auger, 149, 356
Continuous pile wall, 5
Controlled density fill (CDF), 48, 216, 500, 518
Corbels, 85, 89-91
Corner braces, 124, 127-131
Corrosion, 133, 142, 170, 185, 219, 225
Coulomb, 3, 259, 286
Crack tell-tales, 412, 416, 417
Creep, 223, 399, 401-405
ct-Shoring, 495
Cutoff wall, 4, 330
Cylinder pile, 14, 66-69, 175, 223, 235, 268, 345, 352, 495, 499

D

Deadman, 95, 96, 98, 170, 172-174, 314, 499
Debris wall, 240, 243-246
Decking, 17

Deep mixed method (DMM) 5, 48, 49, 356
Deep wells, 330, 331
Design-build, 6
Design load, 395, 399, 402-404
Dewatering, 5, 11, 85, 250, 272, 280, 283, 327, 329-333
Dial gauges, 395, 397
Diesel piledriving, 2, 345, 356, 362
Directional drilling, 7
Drain fabric, 194, 205, 207, 216
Drainage, 9, 194, 199, 223, 225, 233, 272, 341, 342, 426, 427
Drawdown, 330, 332
Drill bit, 4, 149, 150
Drill rig, 345, 352, 358
Drill string, 149
Drilled anchor, 4, 5, 131, 132, 315
Drilled shaft, 3, 59, 102, 317, 506
Drilled soldier pile, 2, 48, 185, 499
Drilled tieback, 4
Driven casing, 4
Driven pile, 4, 131, 317
Driven soldier pile, 6, 12, 48, 49, 85, 307, 317, 499
Dry pack, 80, 81, 85, 86
Drop hammer, 2, 356
Duplex drilling, 4, 72, 149, 363
Dura-Lagg, 199, 201, 202

E

Easement, 9
Earth pressure, 3, 95, 199, 268, 269, 272, 275, 304, 311
Earth pressure coefficient, 250
Earth pressure theory, 3, 267, 268, 276, 279, 322, 499
Earth retention, 1, 3, 5, 6, 7, 25, 27, 225, 259, 261, 393
Elongation, 401, 402, 404-406
Embedment, 175, 304, 307, 308, 312
Encapsulation, 142, 145, 185
Encasement, 317
End bearing, 96
End stops, 71-73
Epoxy coated, 148, 149, 177, 185
Equivalent fluid pressure, 304
Expansive soil, 3

F

Failure plane, 280

Fascia, 51, 66, 71, 73, 74, 78, 79, 177, 185, 187, 188, 194, 199, 206, 224, 230, 232-234, 236, 299, 324, 336, 340
Flocculant, 327
Flushing, 149
Formwork, 14
Foundation excavation, 14
Fracture grouting, 150
Free length, 143, 149, 496
Friction, 95-97, 124, 132, 138, 185, 293, 294, 299, 315-317, 399
Frictional capacity, 9

G

Geo jet, 48, 108, 356, 359-361
Geo support, 1
Geotechnical report, 249, 250, 304
Global positioning, 4, 412
Global stability, 280, 291, 302, 324
Goldberg Zoino Chart, 319, 321
Goldnail, 495, 497, 498
Gow caisson, 2
Gravity, 97, 262, 263
Ground water, 4, 325, 327, 330, 332, 336
Grout valves, 150
Grouting, 5, 124, 126, 131, 132, 136, 138, 142, 149, 150, 170, 175, 177, 185, 194, 199, 201, 206, 258, 316, 356, 361, 363, 372-375, 393, 402, 500, 505, 506, 508
Guaranteed ultimate test strength (GUTS), 402, 403, 501
Guide wall, 71, 72

H

H pile, 39, 89, 131, 161, 165, 499
Haney, 374
Header, 332
Heave, 299
Helical anchors, 131
Hollow stemmed auger, 136, 149
Horizontal drain, 330, 332, 334
Hot rolled, 26
Hutte, 368
Hydration, 363
Hydrofraise, 4, 70, 376
Hydrostatic, 95, 225, 264, 280, 304, 311, 336

I

Impact hammer, 345
In situ test, 3, 240

Interlock, 59, 60, 64
Intermediate piles, 59, 303
Internal friction, 259, 262
Investigations, 249

J

Jack, 41, 92, 102, 134, 279, 398, 400, 460, 490, 494
Jacked piles, 89, 93, 94
Jacking frame, 395

K

K_a, 223, 262, 268-273, 276-279, 286, 304, 307, 308, 311
K_0, 223, 263
K_p, 263, 272, 304, 305, 307, 308, 311
King piles, 59
Klemm, 367, 369, 370
Kobelco, 383

L

Lagging, 2, 6, 14, 15, 17, 33, 45-47, 50-52, 65, 66, 81, 111, 156, 161, 175, 176, 187-194, 196-198, 200-202, 214, 216-218, 220, 223, 224, 226, 230, 231, 235, 245, 261, 268, 299, 307, 317, 319, 322, 327, 336, 376, 411, 412, 495, 499, 500
Lateral earth pressure, 4, 51, 176, 233, 259, 268, 269, 272
Lateral load, 26, 95-97, 124, 126, 225, 230, 267, 268, 279
Lean mix, 48, 59, 63, 345, 500, 519
Limit equilibrium, 3, 267, 280-282, 495, 498
Link-belt, 381
Load cell, 395, 400, 404
Load test, 393
Lookout, 114

M

Manta ray anchor, 131, 132
Manitowac, 380
Mantis, 382
Mass excavation, 12, 14, 80, 81, 214, 216, 217
Mechanical anchor, 131, 132
Mechancially stabilized earth (MSE), 51, 280
Meems, 3
Method of slices, 280, 281, 323
Micropile, 6, 14, 72, 73, 75-79, 89-91, 185, 187, 194, 233, 500

Mohr Circle, 259, 260
Mohr-Coulomb, 260, 261
Moment reduction factor, 311
Monitoring, 4, 407, 408, 412, 415, 418
Movement, 4

N

Nails, 176-178, 182, 185, 199, 206, 230, 233, 299, 323, 324, 402, 403, 498, 507
New Austrian Tunnelling Method (NATM), 7
No-load zone, 132, 138, 142, 280, 314-316, 401, 402, 404, 405, 496

O

OSHA, 419
Overexcavation, 9

P

Paddle mixer, 363
Passive pressure, 170, 263, 268, 273-275, 278, 279, 295304, 307, 308, 314, 315, 318, 496
Passive resistance, 100, 131, 132, 268, 279, 298, 299, 304, 314
Peck, 3, 276, 279
Penetrometer, 3, 261, 441, 444
Perched water, 325, 327
Performance test, 150, 393, 402-404
Permanent anchors, 134, 142, 143, 145-147, 400
Permeation, 3327
Phi, 260
Piezometer, 3
Pile, 51, 89, 92, 103, 109, 114, 118, 151-154, 157, 159, 161, 170, 194, 292-297, 315, 317, 352, 356, 376, 407, 408, 413, 415, 496, 500
Pile driving, 1, 346, 347, 362, 440
Pile splicers, 92
Piling, 1
Pipe, 12, 13, 33, 48, 71, 89, 100, 104, 118, 124, 130, 131, 142, 150, 153, 206
Pipe bedding, 12, 33
Pipe piles, 89
Pit piers, 80, 81, 83
Pneumatic caisson, 2, 5, 6
Polymer, 12
Post tensioned tieback, 4
Post tensioning institute (PTI), 5
Pre-construction survey, 249-257
Pre-existing conditions, 250, 251
Precast, 194, 195, 245

Preloading, 102
Pressure diagram, 276, 277, 279, 304, 305, 307-309, 312, 313, 322, 496
Pressure grouting, 150
Pressuremeter, 3
Primary grouting, 150
Primary piles, 59, 303
Proof test, 150, 393, 404

R

Raker, 4, 95, 96, 99, 100, 102-104, 106, 107, 111-113, 115-117, 216, 291, 292, 294, 295, 297, 317-320, 412, 495, 499
Ram, 395, 399
Rankine, 3, 199, 262, 263, 268, 287, 301, 314
Regrouting, 4
Reinforced earth, 3
Retaining wall, 1, 2, 3, 14, 27, 51, 72, 73, 75, 76, 95, 177, 185, 188, 223-225, 251, 267, 427, 505
Rheology, 5
Road plates, 191, 199
Rock anchors, 131, 142, 170, 219, 225, 315, 393, 403
Rock quality designation (RQD), 250
Roll chock, 109, 111-113, 118, 121

S

Sacrificial anchor, 393,
Sand grout, 150
Scants, 188
Screw anchor, 4
Screw pile, 4
Seal, 325
Secant wall, 5, 14, 16, 17, 59-66, 71, 108, 151, 175, 187, 223, 235, 268, 272, 283, 303, 307, 330, 345, 413, 495, 499
Secondary grouting, 150, 177, 363, 375
Secondary piles, 59
Seismic load, 95, 223
Sewer, 10, 34
Shannon & Wilson, 408
Shear, 97, 132, 199, 240, 259, 261-263, 280, 291, 299, 496
Shear planes, 51
Shear vane test, 250, 444
Sheathing, 136, 138, 142, 185
Sheet piling, 12, 14, 17, 18, 23-29, 32, 64, 65, 100, 118, 124, 170, 172, 187, 222, 223, 227, 235, 240, 268, 272, 283, 298, 303, 307, 308, 311, 327, 330, 345, 411, 495, 496

Sheeting, 1, 421, 422, 459, 460, 475-477
Shoring, 1, 3, 6, 9, 10, 14, 23, 33, 38, 51, 73, 87, 95, 98, 102, 118, 131, 151, 170, 176, 199, 216, 219, 223, 230, 231, 250, 259, 268, 276, 278-280, 283, 287, 291, 295, 303, 306-308, 327, 330, 336, 376, 402, 407, 408, 412, 420-422, 427, 432, 455-457, 474, 475, 479-490, 494-496, 499
Shotcrete, 51, 52, 69, 72, 182, 185, 187, 194, 196, 197, 199, 204, 206, 207, 212-214, 216, 230, 233, 235, 299, 340, 341, 500, 517
Slant pile, 80, 85-88
Slide plane, 52, 66, 240, 301, 332
Slide repair, 47, 235, 240, 241
Slip circle, 300, 301
Slip planes, 51, 274, 301
Slope indicator, 4, 407-412
Slope stability, 5
Slugging test, 250
Slurry, 4, 12, 48, 70, 71, 376
Slurry trench, 4, 330
Slurry wall, 1470-72, 74, 175, 187, 223, 235, 268, 272, 283, 303, 311, 330, 376, 411, 495, 499
Soil anchor, 4, 131, 142, 170, 175, 185, 219, 225, 315, 393, 403
Soil borings, 249
Soil/cement, 5, 48, 51, 330, 356
Soil-Mec, 351, 354, 377, 378
Soil mixed wall, 48
Soil nail, 3, 5, 14, 1751-56, 72, 89, 96, 100, 148, 176, 179-181, 183-185, 187, 199, 203-205, 211, 216, 221, 223, 228, 230, 232-234, 236, 240, 247, 280, 282, 299, 322, 332, 336, 340, 341, 376, 402-404, 410, 495, 497, 499, 500, 507, 508
Soil structure interaction, 259, 268
Soldier pile, 2, 3, 6, 14, 15, 17, 46-52, 65, 66, 81, 84, 85, 95, 96, 98-101, 111, 118, 124, 127, 128, 151, 155, 156, 160-163, 165, 170, 175-177, 187-191, 193, 194, 199, 220, 223, 224, 226, 230, 231, 233, 234, 236-240, 245, 259, 268, 298, 307, 308, 311, 315, 317, 319, 325, 327, 336, 345, 352, 395, 408, 411, 413, 495, 496, 499, 500, 519
Sonic hammer, 2
Spacers, 136-138, 142, 178
Split sets, 177, 180
Stabilizing berm, 52, 57
Standup time, 33, 52, 80, 188, 191, 194, 199, 206, 261, 299

Standard penetration test (STP), 3, 250, 505
Steam hammer, 2
Steel sheet piling, 2, 23
Strain gauges, 3, 407, 412, 415, 418
Strand, 132, 134, 136, 138, 139, 141, 142, 146, 147, 149, 170, 394
Stress, 132, 134, 136, 142, 150, 279, 376, 394, 395, 399, 400-402, 408, 412, 505
Structural concrete, 48, 59, 345, 500
Structural excavation, 12, 13, 27
Stubs, 107, 109, 110, 113, 116-118, 124
Strut, 3, 4, 6, 33, 41, 95, 96, 107, 118-124, 126-128, 185, 276, 278, 279, 291, 294, 308-313, 316, 408, 412, 415, 418
Surcharge, 95, 268.269, 272, 279, 280, 284, 285, 287-289, 446, 475, 495, 496
Surface water, 325, 427

T

Tangent piles, 66, 187, 242, 307, 345, 495
Tendon, 95, 136, 142, 149, 150, 170, 292, 294, 308, 317, 393, 395, 401, 402
Temporary anchors, 134, 399
Terzaghi, 3, 276, 279
Test pit, 249
Texoma, 350
Tic-tac-toe, 199, 204, 206
Tieback, 4, 18, 48, 72, 81, 96, 99, 131, 132, 136, 149-158, 161, 164, 170, 216, 220, 225, 229, 231, 233, 242, 259, 276, 279, 280, 291-294, 308-310, 313-317, 363-366, 368-370, 400, 406, 408, 496, 499, 500, 505, 507, 508
Tiedown, 4
Timber pile, 1, 5
Timber sheet, 23, 24, 39, 40
Timber shoring, 12, 33, 38, 45, 223, 421, 432, 455, 457, 461-472, 489, 494
Top down, 55, 56
Tracked excavator, 214
Trench, 3, 9, 10, 11, 12, 17, 31, 33, 38, 41, 42, 46, 70, 72, 118, 124, 125, 170, 233, 328, 420, 422, 424, 432, 436, 437, 455, 457, 459-473, 475, 476, 483-488, 494, 496
Trench box, 12, 13, 31, 33-36, 187, 219, 223
Trenchless technology, 7, 12
Triaxial tests, 259, 261,
Trumpet, 142
Tunnelling, 12, 17

U

Underpinning, 2, 14, 73, 80-90, 92-94, 187, 223, 422, 427
Uniform soil classification system, 250, 439
Uniformly distributed load (UDL), 268, 272, 283, 287
Unwatering, 5, 326-328, 330
Uplift, 95, 294, 295, 317
Utility location survey, 251, 257

V

Vacuum, 327, 332, 333
Vibro hammer, 2, 4, 51, 99, 345-347, 356
Verification test, 150, 393, 395, 396, 399, 402, 404
Vertical elements, 206, 208-211, 240, 313

W

Wakefield sheeting, 2, 23
Wale, 103, 111, 170, 420, 422, 458-460, 462, 464, 470, 472-478, 482, 485-488
Waler, 40, 100, 102, 103, 107-113, 115-124, 126, 129, 151, 158, 159, 161, 165-172, 199, 203, 206, 292, 295, 412, 475, 482, 485-488, 499
Water control, 325
Water cutoff, 9, 11, 12, 71
Water table, 27, 51, 250, 272, 325, 327, 330, 332, 496
Waterstop, 71, 73
Watson, 348, 353, 356, 366
Wedge theory, 3
Wedge, 107, 118, 124, 136, 161, 188, 262, 263, 301, 314
Weir, 327
Wellpoint, 330, 333,
Wet set, 48, 51, 149

Y

Young's Modulus, 402